THE NEW NATURALIST
A SURVEY OF BRITISH NATURAL HISTORY

WILD FLOWERS

The aim of this series is to interest the general reader in the wild life of Britain by recapturing the inquiring spirit of the old naturalists. The Editors believe that the natural pride of the British public in the native flora and fauna, to which must be added concern for their conservation, is best fostered by maintaining a high standard of accuracy combined with clarity of exposition in presenting the results of modern scientific research. The plants and animals are described in relation to their home and habitats and are portrayed in the full beauty of their natural colours, by the latest methods of colour photography and reproduction.

THE NEW NATURALIST

WILD FLOWERS

Botanising in Britain

by

JOHN GILMOUR

M.A., V.M.H.

DIRECTOR OF THE UNIVERSITY BOTANIC GARDEN
AND FELLOW OF CLARE COLLEGE, CAMBRIDGE

and

MAX WALTERS

M.A., Ph.D.

CURATOR OF THE UNIVERSITY HERBARIUM AND
FELLOW OF KING'S COLLEGE, CAMBRIDGE

WITH 45 COLOUR PHOTOGRAPHS
27 PHOTOGRAPHS IN BLACK AND WHITE
AND 3 LINE DRAWINGS

COLLINS
ST JAMES'S PLACE LONDON

To
Molly and Lorna

02282996

NELSON

ISBN 0 00 213251 6

First Edition: 1954
Second Edition: 1955
Third Edition: 1962
Fourth Edition: 1969
Fifth Edition: 1973

PRINTED IN GREAT BRITAIN
COLLINS CLEAR-TYPE PRESS: LONDON AND GLASGOW

CONTENTS

	PAGE
Editors' Preface	xi
Authors' Preface	xiii

CHAPTER

1	The Anatomy of Field Botany	1
2	How our Flora was Discovered	7
3	The Biology of our Flora	28
4	How our Flora Came to Britain	47
5	Woodlands and Hedgerows	57
6	Moors, Heaths and Commons	81
7	Chalk Downs and Limestone Uplands	96
8	Mountains	113
9	Bogs, Fens and Marshes	131
10	Rivers, Lakes and Ponds	148
11	The Sea Coast	166
12	Fields and Roadsides	182
13	Arable Land, Waste Ground and Walls	189
14	How to Go Further	200
	Appendix I: Ways and Means	205
	Appendix II: Keys	211
	Bibliography	224
	Index-Glossary	229

PLATES IN COLOUR

		FACING PAGE
1	Goatsbeard, *Tragopogon pratensis*, in fruit	38
2a	Bugle, *Ajuga reptans*	39
2b	Betony, *Stachys officinalis*	39
3	Lords-and-Ladies, *Arum maculatum*	46
4	Bluebells, *Endymion nonscriptus*	47
5	Yellow Dead-nettle, *Galeobdolon luteum*	50
6	Lesser Celandine, *Ranunculus ficaria*	51
7a	Wood Sorrel, *Oxalis acetosella*	66
7b	Winter Aconite, *Eranthis hiemalis*	66
8a	Wood Anemone, *Anemone nemorosa*	67
8b	Common Violet, *Viola riviniana*	67
9	Broom, *Sarothamnus scoparius*	82
10a	Gorse, *Ulex europaeus*	83
10b	Cotton-grass, *Eriophorum vaginatum*	83
11	Heather, *Calluna vulgaris*, and Scots Pine, *Pinus sylvestris*	98
12a	Thyme, *Thymus drucei*	99
12b	Bird's-foot Trefoil, *Lotus corniculatus*	99
13	Wild Pansy, *Viola tricolor*	114
14	Meadow Cranesbill, *Geranium pratense*	115
15	Alpine Lady's Mantle, *Alchemilla alpina*	130
16a	Moss Campion, *Silene acaulis*	131
16b	Starry Saxifrage, *Saxifraga stellaris*	131
17a	Bladderwort, *Utricularia vulgaris*	146
17b	A Marsh Orchid, *Orchis strictifolia*, and Marsh Horsetail, *Equisetum palustre*	146

FACING PAGE

18*a* White Water-lily, *Nymphaea alba*   147

18*b* Arrowhead, *Sagittaria sagittifolia*   147

19   Yellow Flags, *Iris pseudacorus*   162

20   Grass of Parnassus, *Parnassia palustris*   163

21   Felwort, *Gentianella amarella*   164

22   Sea Lavender, *Limonium vulgare*   165

23*a* Sea Pink, *Armeria maritima*   176

23*b* Biting Stonecrop, *Sedum acre*   176

24   Meadowsweet, *Filipendula ulmaria*   177

25*a* Cowslip, *Primula veris*, and Early Purple Orchid, *Orchis
       mascula*   178

25*b* Wild Daffodils, *Narcissus pseudonarcissus*   178

26*a* Mullein, *Verbascum thapsus*   179

26*b* Yellow Toadflax, *Linaria vulgaris*   179

27*a* Scentless Mayweed, *Matricaria maritima* subsp. *inodora*   182

27*b* Bindweed, *Convolvulus arvensis*   182

28*a* Poppy, *Papaver rhoeas*   183

28*b* Field of Buttercups, *Ranunculus* spp.   183

29   Deadly Nightshade, *Atropa belladonna*   190

30   Coltsfoot, *Tussilago farfara*   191

31   Welsh Poppy, *Meconopsis cambrica*   194

32   Rose Bay Willow-herb, *Chamaenerion angustifolium*, and
       Oxford Ragwort, *Senecio squalidus*   195

*It should be noted that throughout this book Plate numbers in arabic figures
refer to the Colour Plates, while roman numerals are used for
Black-and-White Plates.*

PLATES IN BLACK AND WHITE

FACING PAGE

I Field botanists in action   18

II *Carex pendula*   19

III Stinking Hellebore, *Helleborus foetidus*   34

IV Herb Paris, *Paris quadrifolia*, with Dog's Mercury, *Mercurialis perennis*   35

V Marjoram, *Origanum vulgare*   70

VIa Centaury, *Centaurium erythraea*   71

VIb Wild Strawberry, *Fragaria vesca*   71

VII Maiden Pink, *Dianthus deltoides*   78

VIII *Lobelia urens*   79

IX Spiked Speedwell, *Veronica spicata*   102

X Roseroot, *Sedum rosea*   103

XI The Sundew *Drosera anglica*   110

XII Pale Butterwort, *Pinguicula lusitanica*   111

XIII Water Germander, *Teucrium scordium*   134

XIV Bog Bean, *Menyanthes trifoliata*   135

XV Sweet Flag, *Acorus calamus*   142

XVI Lesser Water-plantain, *Baldellia ranunculoides*   143

XVII Mud Crowfoot, *Ranunculus lutarius*   150

XVIII Lupins, *Lupinus nootkatensis*   151

XIX Marsh Helleborine, *Epipactis palustris*   158

XX Young dunes with Marram Grass, *Ammophila arenaria*   159

XXIa Sea Pea, *Lathyrus japonicus*   166

XXIb Strawberry Clover, *Trifolium fragiferum*   166

FACING PAGE

XXII   Lesser Broomrape, *Orobanche minor*   167

XXIIIa The Speedwell *Veronica praecox*   174

XXIIIb Overgrown steps of deserted house   174

XXIV   Making a herbarium specimen   175

EDITORS' PREFACE TO THE FIRST EDITION

THE MODESTLY EXPRESSED intention of the authors of this book is to introduce British wild flowers to those who, though keen and interested, may feel in need of help and guidance.

We feel that this first book in the *New Naturalist* series by our botanical editor—in happy partnership with his able and learned colleague, Dr. Walters—succeeds in being far more than is suggested by this. With its accent upon history, ecology and the communities of plants, WILD FLOWERS succeeds, in our view, in making a really important contribution to the understanding of British plants by the general naturalist, and by those working in other fields of natural history. It is just the sort of book that the *New Naturalist* was made for. It is a fitting companion to Dr. Turrill's BRITISH PLANT LIFE, and in conjunction with it forms a valuable basis for appreciating the other botanical volumes in the series, which deal with more restricted aspects of the subject.

One of the qualities of WILD FLOWERS is the successful way in which it recaptures the sense of wonder and freshness felt by the naturalist who has newly become a naturalist, whether he be young or old—the excitements, the difficulties, the mistakes and the triumphs of the beginner; we are fortunate that these are things which the authors have not been able to forget. Both were trained at Cambridge and have since become professional botanists—John Gilmour as Assistant Director at the Royal Botanic Gardens at Kew, later as Director of the Royal Horticultural Society's Gardens at Wisley, and now as Director of the University Botanic Garden at Cambridge; and Max Walters as Curator of the Herbarium at Cambridge and as a leading research worker in experimental taxonomy.

The personal histories of our two authors have thus qualified them well to greet the new recruits to botany for whom this book is primarily designed. Both have experience and expertness not only in field botany and research, but in teaching; and both have had a close connection with the Botanical Society of the British Isles, of which John Gilmour was President for several years. Throughout this volume the writers have stressed the part that amateurs can play, by observation, record and experiment, in building up a more complete knowledge of British plants. In the world of natural history the importance of the amateur as fact-finder, research worker and intelligencer has never been greater than to-day, and the further mobilisation of amateur help would be of particular service to the development of British field botany.

We particularly commend the felicitous historical approach of our authors to both botany and botanists. We suspect that the chapter *How Our Flora was Discovered* is the only available account of the history of British field botany, and we must confess to a certain editorial disappointment at its briefness while at the same time understanding why it has to be so brief when there is so much else to say. Among other things that we have found delightful are the numerous, unfamiliar but always relevant quotations from English poets, so many of whom seem to have been amateur botanists, and good ones too.

Field botany to-day is composed of several interweaving strands. For naturalists the traditional study of identification and distribution must come first. (We hope that the keys our authors provide to the yellow-flowered composites and the hedgerow umbellifers will smooth the path of beginners tackling these twin bogeys for the first time.) But our authors lead us beyond identification to modern knowledge of the structure and history of plant communities and the ecological preferences of individual species; and to the more recent discipline of experimental taxonomy, which seeks to unravel the evolutionary significance of the units composing a particular flora.

It is because this volume forms an introduction to all aspects of field botany, new as well as old, that we are confident that it will help many to take a broader and deeper view of what is meant by its sub-title: *Botanising in Britain.*

THE EDITORS

AUTHORS' PREFACE TO THE FIRST EDITION

OUR AIM in writing this book has been to provide an introduction to the British flora for those who are keen on wild flowers, but who feel that they want some help and guidance in their keenness. It is designed, not as a reference book for the field or the study, but as a volume for leisurely and intermittent reading in an arm-chair, a bed or a train. Although we have included one or two keys to groups difficult for beginners (such as the hedgerow umbellifers and the yellow " dandelion composites ") the book is not intended for the identification of species. Many excellent volumes already exist for this purpose, from elementary pocket-books to the full-scale *Flora of the British Isles* by Clapham, Tutin, and Warburg.

After four introductory chapters, providing a background of knowledge on the biology and history of the British flora and how it has been discovered, together with some notes on the " anatomy " of botanising as a human activity, the main part of the book consists of a description of the most important British habitats and the characteristic plants that they contain. Here we have obviously had to make a rigorous selection of species. Our aim has not been to provide a catalogue of the British flora, but to concentrate on those species which are either important constituents of the community concerned or which illustrate points of general biological interest. Many common plants have perforce been omitted and a number of rare species included for some special reason. The last chapter suggests ways in which a keen student may take his studies further, and links up with Dr. Turrill's more advanced book in this series, *British Plant Life* (1948). There is an appendix dealing with the equipment necessary for collecting specimens and making a herbarium, and with the books available on various aspects of field botany. A single bibliography at the end of the volume gives fuller particulars of the brief references scattered throughout the text, and there is a glossary combined with an index.

The Latin names used are, with few exceptions, those in Clapham, Tutin and Warburg (Ed. 2, 1962); some of their better-known synonyms have been included in the index, with a cross-reference to the *Flora* name.

A word must be said about the joint authorship of the book. In the actual writing, as will be seen from the initials, each author is responsible for specific chapters or sections, though all chapters have been revised and discussed by both authors. We realise that this method leads to a certain disparity of style, but we hope that this will prove a mild stimulant rather than an irritant to our readers. We chose this method as it enables personal experiences and impressions to be recorded, and minimises the use of the editorial "we."

We should like to record our grateful thanks to the Editors and Publishers of the series, and to Mr. J. E. Raven, Mr. J. E. Lousley, and to Sir Julian Huxley and the late Mr. James Fisher for many valuable suggestions. We would also like to thank Mr. F. E. Stoakley for assistance with the bibliography.

Note on the Fifth Edition

In this edition we have, as far as practicable without complete re-setting, incorporated the alterations and additions included in the recently published Fontana paper-back edition; in particular, the Bibliography and the Book section of Appendix I have been brought up to date, omitting several older references and substituting a number of more modern ones.

J. G.
M. W.

CHAPTER I

THE ANATOMY OF FIELD BOTANY

(J. G.)

THE FIRST STEP towards becoming a field botanist—as towards becoming a stamp collector, a rock climber or a prima donna—is to be visited by an irresistible passion. Like Shelley's "Spirit of Beauty," such passions are unpredictable in their comings and goings.

> " *When musing deeply on the lot*
> *Of life, at that sweet time when winds are wooing*
> *All vital things that wake to bring*
> *News of birds and blossoming,—*
> *Sudden, thy shadow fell on me;*
> *I shrieked, and clasped my hand in ecstasy!* "
> (*Hymn to Intellectual Beauty*)

Without an experience of the shadow's fall and an answering shriek of ecstasy no one has ever become a field botanist. The great seventeenth-century naturalist John Ray described his own awakening in the following words: " First the rich array of spring-time meadows, then the shape, colour and structure of particular plants fascinated and absorbed me: interest in botany became a passion." (Raven, 1950, p. 81.)

This essential passion may develop, understandably enough, from the stimulus of family or friends, but as often, perhaps, it arrives unheralded and inexplicable—even against all reasonable probabilities, as it did, if I may be forgiven a little autobiography, in my own case. When I went to a preparatory school I knew and cared nothing about wild plants. At the end of the summer term each boy had to produce fifty named species. On the last day but one I had not collected a single plant. Desperation drove me to a high-speed tour of the lanes near the school, guided by a friend who had already made his collection, and on the following day I duly presented my fifty plants. This discreditable incident implanted in me, against every modern principle of education, a passion for the British flora which has never been extinguished. The following summer holidays I spent nearly every waking moment combing the Devonshire hedgerows, carrying Johns' *Flowers of the Field* in one hand and a shining new vasculum in the other.

However it may come, the passion must be there; but passion alone is, of course, not enough. Bertrand Russell has defined the " good life " as one that is " inspired by love and guided by knowledge." In Appendix I, at the end of the book, we have tried to set out in detail some of the basic methods of acquiring and recording the " guiding knowledge " necessary for a well-equipped field botanist; in the present introductory chapter I want to consider, very briefly, what might be called the " anatomy " of field botany as a labour of love and the effect of the personal factor on botanising aims and methods.

In the first place, then, why do we botanise? What are the components of the passion that urges on its victims? The enjoyment of the beauty and fascination of flowers in their natural settings, the excitement of personal discovery, and the satisfaction of the collecting " instinct " are probably always present, but if with these can be blended the aim of adding something new, however small, to the general body of knowledge of our flora, a fresh and stimulating enjoyment can be felt. Such a stimulus will lead us on where more private exaltation may fail or falter, while the new fact discovered, however apparently trivial, will, if properly recorded, be of real value to the scientific study of British botany. As we have emphasised throughout this book, even in so well-botanised a country as ours, there is a great deal still to be observed and recorded about the plants it contains, their distribution, their life-histories, and their habitats. I do not suggest, of course, that a beginnner can expect to make such discoveries as soon as he starts botanising, but, once he has found his botanical feet, his steps should be lured on by the possibility of new scientific knowledge awaiting him round the other side of the wood.

There are several avenues along which this quest for new botanical knowledge can be pursued. For example, a study can be made of the plants in a limited area, say a parish, a county, or a river valley, leading, often, to the writing of a local Flora. This appeals strongly to many people, as it brings in the emotion of local patriotism. Alternatively, a particular group of plants can be chosen and investigated as thoroughly as possible. This second type of work can be done from what may be called a purely systematic angle, that is to say the plants identified as accurately as possible and their distribution recorded, or it may be tackled from a more broadly biological and ecological point of view. The relationship between these two approaches is more fully discussed in the last chapter of the book.

Whichever approach is preferred, it cannot be too strongly emphasised that good work in the field must be the basis of all worthwhile botanical enterprise. Details of equipment are described in Appendix I, but the technique of planning and carrying out a collecting expedition is equally important and is as varied as the temperaments of botanists. The primary division is between those who need a crowd, those who prefer a maximum of one or two carefully chosen friends, and those to whom any companion is a desecration and a torture. The middle course normally yields the most satisfactory results and—unless temperament irresistibly forbids—is to be recommended. Whether alone or in company, however, it is wise always to have a specific object for the expedition. This may seem an unnecessary cramping of the wayward fancy usually associated with the " Nature Lover," but all experienced field botanists would agree that it pays in the end—in both results and enjoyment. The actual object will, of course, vary according to the stage reached by the botanisers. For those whom the passion has just visited the aim may be to try and disentangle some, at any rate, of the puzzling yellow composites that look so alike (the hawkbits, hawk's-beards, hawkweeds and their relations—see p. 218), or to be introduced to some uncommon plant for the first time. More advanced addicts may have in mind the study of a difficult group in which they are specialising, say the eyebrights or the marsh orchids, or, if they are preparing a local Flora, they may visit a little-known spot in their area to make lists of the common species growing there, with the ever-present hope, at the same time, of lighting on an unsuspected and unrecorded rarity.

Unless you are an incurable solitary, you should join, as soon as possible, your local natural history society, if one exists. If it does not, try to form one. A high proportion of the field botany done in this country during the last hundred years has been under the auspices of such societies, and they provide an indispensable rallying point for the production of local Floras and other co-operative undertakings. Some of the larger societies produce journals in which local records can be published. A very useful list of local societies has been compiled by Lysaght (1959).

In addition to local bodies, there are two main national societies concerned with field botany, the Wild Flower Society and the Botanical Society of the British Isles (address: British Museum (Natural History), London, S.W.7). The first is for beginners, while members of the second range from the slightly more advanced to the professional expert. The journal of the B.S.B.I. is christened *Watsonia*, after H. C. Watson (see p. 24); it is now the main vehicle for the publication of papers on British field botany. In addition to its journal, the B.S.B.I. organises a full programme of summer field excursions, as well as conferences and exhibition meetings. There are local secretaries and recorders all over the country, and a panel of experts who identify specimens and provide other information for members. Every serious field botanist should join this society. Since 1954, B.S.B.I. members have been participating, along with many other amateur and professional botanists, in a detailed recording survey of the distribution of British plants. The Society's Distribution Maps Scheme has been outstandingly successful, and well over $1\frac{1}{4}$ million records were collected and used in preparing the maps for the *Atlas of the British Flora* (Perring and Walters, 1962; Perring and Sell, 1968).

Planned botanising is undoubtedly the right policy. Nevertheless, the unplanned variety can often be secretly indulged during an hour's wait after a missed bus, or on a family outing—and is sometimes surprisingly productive. The classic example in recent years is Mr. J. E. Lousley's re-discovery of the military orchid on the Chilterns during a family picnic—though Mr. Lousley, in his account of the affair (1950, p. 89), says that the hamper was opened in a spot selected by himself, rather than by his family, and no doubt his unrivalled knowledge of the orchid and its past occurrences played a part in his choice. In the re-discovery of *Orchis militaris*, serendipity, one suspects, was nicely balanced by hope many times deferred.

Before saying a final word about keeping records of the results of field work, the opportunity must be taken to sound a warning—often repeated but still, unfortunately, too often neglected—against the over-collecting of uncommon species. This is a matter, not of reason, but of emotion. The rational arguments against taking specimens from a small colony of a rare plant are obvious and overwhelming, but every collector knows that, when a long-desired specimen lies at his feet, the voice of reason is apt to grow faint. The answer, of course, is to fight emotion with emotion. The love and respect for our heritage of wild life must be so strong that to damage or diminish it becomes unthinkable. This outlook can be achieved only by education, and many societies, schools and other bodies have been doing excellent work in this field during the past forty or fifty years. It is good news that the B.S.B.I. has adopted a "Code of Conduct" for its members; a leaflet (1970) setting out a sensible policy on collecting embodied in this code is being made widely available, and can be obtained from the Society. If your collector's instinct is so powerful that it cannot tolerate gaps in your collection, the answer is undoubtedly to be found in photography, drawing or painting. As with the hunting of big game in Africa, the modern trend in plant collecting is to rely on a first-class picture and leave the rare beauty for others to admire.

In recent years naturalists have become far more aware of their responsibility to safeguard their rich heritage, and the vigorous growth of County Naturalists' Trusts, supported by the national Society for the Promotion of Nature Reserves, is a healthy sign. Local natural history societies and trusts are now co-ordinated in a Council for Nature, one of whose most successful " children " is the Conservation Corps, for young people to take part in work camps on nature reserves. The government organisation known as the Nature Conservancy, set up in 1949, has created a system of National Nature Reserves, and is concerned with much research on conservation. An authoritative account of the whole field of nature conservation in Britain was written for this series by the late Sir Dudley Stamp (1969).

Records of field work are of two kinds, written records and dried herbarium specimens. As to the first, it is clearly foolish to lay down hard and fast rules; everything depends on temperament and outlook. Some may be natural card-index compilers and in the evening of their days be able to point lovingly to the card recording their first find at the age of eleven; others may spurn such mechanical aids and prefer to rely on a scribbled note in the margin of a Flora. The golden rule is— *write it down at once.* There is no need to labour this point; the evanescence of human memory is too frequent and bitter an experience. In order to put this admirable precept into practice, the essentials are a note-book for carrying in the field (a loose-leaf one with stiff covers is ideal), and some form of more permanent and easily consulted record for use at home. The B.S.B.I.'s Maps Scheme record cards have revolutionised field recording for many botanists; they will undoubtedly continue to be used in the compiling of County Floras long after the main work of the Maps Scheme has been completed. I am a great believer in card-indexes. Despite their bulk for transport, the ease of arrangement, rearrangement, and cross-referencing gives them, on balance, an advantage over note-books. Even more important, of course, than the mechanics of record-keeping, is what is put into the records and how it is arranged. Here it all depends on the object of the botanising. Clearly, if you are writing a Flora of the bombed sites of Coventry, the arrangement and content of your records will not be on the same plan as if you are studying the distribution of brambles in Southern England. Whatever your object, however, make your notes as full as possible and always carry in your mind all the possible future uses of the records. There is nothing more annoying, when you finally come to put the results on a piece of work together, than to be forced to utter the *cri du coeur*—" If *only* I had made a note of that."

Most serious field botanists make a herbarium, though it is not essential to the enjoyment of their hobby, nor their proficiency in it. There are two main reasons for preserving dried plants, one " external," as it were, and the other " internal." The first is, of course, to create a permanent record that can be re-examined, and challenged or confirmed, by later workers. This is the rational justification, but it is doubtful whether so many private herbaria would be formed were it not for the gratification it gives to the collecting urge so deeply rooted in man (though less developed in woman!). Some good herbaria have been created almost entirely for this " internal " reason, but collections are likely to be better documented and arranged if the " external " reason is also present. If, as I have urged, permanent addition to knowledge is to be a main aim of botanising, the formation of a herbarium is, of course, essential. Such a herbarium may be kept as the private property of its maker, or the sheets may be deposited after a time—or as they are made—at a public institute such as a local museum. Indeed, it is frequently urged that the latter course is the right one, as the plants are then available for all to consult.

The passion for field botany, then, may take many forms and work itself out in many ways. What is certain is that those who have been visited by it will defend the claims of field botany against all rivals as an activity that takes them into pleasant places, in company with pleasant people, and contributes not a little to our knowledge of the world around us.

CHAPTER 2

HOW OUR FLORA WAS DISCOVERED
(J. G.)

THE TWO THOUSAND-ODD species of flowering plants found wild in the British Isles have probably been more thoroughly studied than has the flora of any other country.   The story of the gradual building up of this knowledge begins in the far-off days of the Celtic and Saxon herb-gatherers, but it was not until the middle of the sixteenth century, when the great wave of Renaissance learning reached our shores, that the crude lore of the medicine-man gave way to the first groping explorations of the scientist.

In the 1520's, at Pembroke College, Cambridge, a young medical student was in the habit of playing tennis with his friends Hugh Latimer and Nicholas Ridley.   He was William Turner, later Dean of Wells, fiery Protestant controversialist and the " Father of British Botany."   Turner was the first in Britain to light his torch at the flame of the pioneer herbalists on the Continent, and, breaking away from authority and superstition, to describe British plants from his own observation and experience.   Among the three hundred native species he mentions in his various works, were yellow flag from his native Northumberland, dog's mercury from Trinity College, Cambridge, and bluebells " muche aboute Sion," perhaps the ancestors of those now carpeting the Queen's Cottage grounds opposite Syon House at Kew.

Like Turner, the majority of sixteenth-century botanists were medical men.   Plants were then the chief weapon in the doctor's armoury, and botany did not emerge as an independent science until well into the seventeenth century.

Two or three years before Turner's death in 1568, a Flemish doctor, Mathias de L'Obel, settled in England.   Like many another such visitor he made an outstanding contribution to the knowledge of our island. Several years' intensive botanising—including an exploration of that now famous Bristol locality, St. Vincent's Rocks—enabled him to publish records of eighty plants unknown to Turner. The flowering rush he found near the Tower of London, and the lovely yellow-flowered floating *Nymphoides peltatum* " juxta amoenissima Thamesis fluenta," where it still grows. L'Obel became Botanist to James I, and continued his labours for British botany until his death at Highgate in 1616; he is commemorated in the name *Lobelia*.

John Gerard, author of the famous *Herball* (1597), is in ill repute as a shameless plagiarist, having, without acknowledgement, based that ever-popular book on the manuscript of another, but his explorations of the British flora were original and outstanding. We can picture him setting out from his Physic Garden at Holborn into the fields and woods to the north of London, and carrying back, in some Elizabethan ancestor of the modern vasculum, yellow archangel and twayblade from Hampstead and adder's tongue fern from " the Mantells." In his *Herball* the number of British species listed had risen to about 500, of which 182 were first records for Britain.

On 13 July 1629 a company of men met at St. Paul's, walked to the river, and set sail for Gravesend and the Kentish countryside. Though their leader, Thomas Johnson, was an apothecary and most of the others were medical men, the publication later in the year of Johnson's *Iter . . . in agrum Cantianum* describing their adventures and the plants they found, was a landmark in British field botany, since the book deals with plants primarily as plants and not as herbs for medical use.

Johnson's party embarked in two boats. Shortly afterwards a storm broke and one of them had to run into Greenwich for repairs. Supper at the Bull at Rochester, however, saw the party united again and busy making a list of the plants in the inn garden. Their route during the remainder of the journey lay through Chatham, Gillingham, Queenborough (where the Mayor began by demanding an explanation of their presence and ended by treating them to some of his best ale) and the Isle of Grain, and so back by Dartford to Erith, where they embarked on an East Indiaman for London Bridge, arriving on 17 July. During the return journey Johnson showed his mettle by scorning the services of the wagon into which all but one of his weary companions were glad to climb on arrival at Stoke. Altogether the party recorded 250 plants during their five-days' journey.

As an Appendix to the *Iter* Johnson printed an account of an expedition to Hampstead and its environs, undertaken by practically the same party on 1 August. Three years later, in 1632, he gathered together all existing records for the Hampstead neighbourhood into the first printed " Local Flora " to be published in Britain (Johnson, 1632, after the Kentish section), and the forerunner of a vast corpus of such works now covering nearly the whole of the kingdom. Hampstead Heath has been continuously botanised since the days of Gerard and Johnson, and it must now be botanically one of the best documented localities in the world.

Johnson's Kentish Journey in 1629 was the prelude to others of the same kind—through Kent again in 1632 and later farther afield in southern England and in north Wales. His avowed aim was to publish a Flora of the whole of Britain; and the first two volumes of his *Mercurius Botanicus* (1634 and 1641), containing a list of all the native species then known to him (about 700, of which nearly 170 Johnson himself had recorded for the first time in Britain), can be regarded as a prelude to the more ambitious work. But it was not to be. In 1644, at the age of 40, he was killed while fighting for the King at the siege of Basing House. Johnson's *Iter* (1629), *Descriptio* (1632), and the two parts of the *Mercurius* (1634 and 1641), which are all very rare books, were reprinted in 1847. They were written in Latin, but English translations of the *Iter* and the *Descriptio*, by the late Canon Raven, together with facsimiles of the first editions, were published in 1972 by the Hunt Institute of Botanical Documentation at Pittsburgh, U.S.A., as the third volume of their Facsimile Series, under the title *Thomas Johnson—Botanical Journeys in Kent & Hampstead*.

Johnson's scorn of the wagon at Stoke in 1629 was prophetic of the indomitable energy and courage he was to infuse into all that he undertook. Into his forty years he crammed a prodigious output of work (he edited and vastly improved Gerard's *Herball*, 1633, and translated the works of Ambrose Parey, the great French surgeon), and won the admiration of all for his bravery during the Civil War. After his death at Basing House he was described as " no lesse eminent in the Garrison for his valour and conduct as a souldier, than famous through the kingdom for his excellency as an Herbarist and Physician."

In certain Scottish pinewoods there grows the little orchid *Goodyera repens*, named by Robert Brown in memory of John Goodyer, Johnson's constant helper and closest botanical friend. Goodyer was a man who preferred to give his work to the world through others rather than to publish himself, and his outstanding merits as a field botanist were not fully realised until Dr. Gunther studied his library and papers at Magdalen College, Oxford. In some ways Goodyer was a more striking figure than Johnson. He was an amateur botanist, employed as steward to Sir Thomas Bilson at Mapledurham in Hampshire, and his interest in plants was quite uncoloured by medical necessity. On horseback he visited many parts of Southern England, and Johnson's edition of Gerard was greatly enriched by his discoveries. Johnson described Goodyer as " a man second to none in his industrie and searching after plants nor in his judgment or knowledge of them," and he was Johnson's chosen collaborator for the British Flora that was never written.

In 1629, the year of Johnson's *Iter*, there appeared John Parkinson's *Paradisi in Sole Paradisus Terrestris* (Park-in-Sun's Park on Earth), prefaced with commendatory verses by Johnson and others. Parkinson (1567-1650) was mainly a horticulturist, but in his *Herball* (1640) he showed that he was also a good field botanist, recording twenty-eight species for the first time in Britain, including two famous plants—the strawberry tree (*Arbutus unedo*) from " the West part of Ireland," and the lady's slipper orchid (*Cypripedium calceolus*) from Lancashire.

Johnson gathered many helpers around him, notably the Rev. Walter Stonehouse, a Fellow of Magdalen College, Oxford, who was with Johnson in north Wales in 1639, and George Bowles of Chichester. During the second quarter of the seventeenth century Johnson was indeed the key figure in British field botany, consolidating the work of the early herbalists and preparing the ground for the great work of John Ray which was to come.

Before, however, we shift the scene to Ray's rooms at Trinity College, Cambridge, two books must be mentioned whose titles— *Phytologia Britannica*, and *Pinax rerum naturalium Britannicarum*—are grandiloquent, but which in fact signify less than one is led to expect. William How's *Phytologia* (1650) was a reasonably well compiled list of previously printed records (based largely on Johnson's *Mercurius*), together with a few new records supplied by friends of the author, such as Stonehouse, but it was nothing more. How was a busy physician who had no time for scouring the countryside for new plants, and he did not make any original contributions himself. The *Phytologia* was long regarded as the first British Flora, but, as shown by Kew and Powell (1932), this honour should more properly go to Johnson's

Mercurius, which, in substance though not in form, was a complete list of the then known British flora. In Goodyer's library at Magdalen College, Oxford, is Johnson's own copy of his *Descriptio* (1632) with How's manuscript notes for the *Phytologia* written on blank pages at the end; also How's copy of the *Phytologia* with manuscript additions for a second edition which he did not live to publish. Goodyer's papers and books at Magdalen, fully described by Gunther (1922), form a wonderful storehouse of first-hand information on early British field botanists; deciphering his crabbed seventeenth-century script one can live again the excitement of the pioneer work on our flora carried out by Goodyer, Johnson and their learned and ingenious friends.

Christopher Merret's *Pinax* was a disappointment to Ray when it appeared in 1666 ("the world," he said, "is glutted with Merret's bungling *Pinax*"), and we have no reason now to quarrel with his judgment. Merret was a member of the Johnson-Goodyer circle, a London physician and one of the early Fellows of the Royal Society; but his knowledge of botany was superficial. The *Pinax* was an uncritical compilation from previous writers, with between 40 and 50 new records, including some of his own observing, and some sent by Thomas Willisel, an old soldier, who was later employed as a collector by Ray. Only about 400 native species were mentioned, though W. A. Clarke (1900) estimates that, at that time, records of over 700 species had been published. The *Pinax* was reprinted in 1667, copies dated 1666 being extremely rare, through being destroyed in the Fire of London.

The stage was now set for the appearance of the greatest figure in the story of British field botany. Johnson and Goodyer had failed to produce a full-dress descriptive Flora, but it was their unfinished labours which formed a background for John Ray, blacksmith's son and dissenting divine, to carry through triumphantly a task which they could only plan.

Ray was born at Black Notley, near Braintree, Essex, in 1627, two years before Johnson published his first book. The cottage behind the forge, which tradition says was his birthplace, is still standing, and, in another part of the village, outhouses mark the site of the house which Ray built for his widowed mother in 1655, and in which he died in 1704. Dr. Raven, in his masterly book on Ray (1950), has traced the course of his life—and expounded his work—so minutely and sympathetically that any subsequent writer can only select gratefully from the wealth of material now to his hand. Here I must reluctantly give only the barest outline of Ray's life, ignore completely his outstanding contributions to zoology, folk-lore and other sciences, and, even within the sphere of botany, confine myself strictly to his work on the British flora.

Through the interest of the Vicar of Braintree, Ray went up to Catharine Hall, Cambridge, in 1644, the year that Johnson died of his wounds at Basing House. Two years later he transferred to Trinity, where he became Fellow (1649) and Tutor (1653). He was ordained in 1660, but in 1662 he resigned his appointments at Trinity owing to his refusal to assent to the Act of Uniformity. Dr. Raven has given a vivid picture of the struggle in Ray's Cambridge between the traditional disciplines of grammar, logic and rhetoric, with their public disputations and " opponencies," and the awakening interest in " natural philosophy " that was to revolutionise the world.

In this struggle Ray played an important, though not, at the time, spectacular part. He belonged to a small circle of friends, who, amidst the indifference, ridicule or opposition of their contemporaries, carried out experimental and observational work in various branches of science, notably comparative anatomy and chemistry. In this atmosphere Ray's love of botany first took possession of him. During his recovery from an illness he began to explore the country round Cambridge, and later to grow wild plants in the little garden attached to his rooms at Trinity—quite possibly, Dr. Raven thinks, the same rooms which, nearly forty years after, were occupied by Isaac Newton.

The explorations of Ray and his friends, of whom the chief was John Nidd, led in 1660 to the publication of Ray's *Catalogus Plantarum circa Cantabrigiam nascentium*. This was the first work devoted entirely to an enumeration of the plants of a limited area in Britain, but it was something much more than this.

" Few books of such compass," writes Dr. Raven, " have contained so great a store of information and learning or exerted so great an influence upon the future; no book has so evidently initiated a new era in British botany." What did Ray do to justify such eulogy? His chief service, unaided except by his own genius and industry, was to bring some sort of order out of the chaos of synonyms, cross-references, and obscure descriptions which by then filled the works of his predecessors, Gerard, Parkinson, Johnson and the rest. Looking back, we can see Ray's pocket Catalogue as a sort of sieve into which he threw the rich confusion of the previous hundred years' work, and from which he shook, not indeed a pure product, but at any rate a manageable mixture of names on which he and his successors were able to build a sound knowledge of our flora. The Cambridge Catalogue is undoubtedly one of the keystones of our botanical structure; it is also a charming—and amazingly comprehensive—account of the plants of the prolific Cambridge countryside. Unfortunately for present-day readers, but fortunately for Ray's international reputation among his contemporaries, it is written in Latin. Nevertheless, the effort (mitigated by the English description of the localities) of dipping into it is well worth while. Of the approximately 950 species listed in Babington's *Flora of Cambridgeshire* (1860), over 700 were recorded by Ray. Very few of the Cambridge treasures escaped his eye, and the field botanist who to-day finds *Linum anglicum* and *Thesium humifusum* on the Gogs, *Melampyrum cristatum* in Hardwick Wood, and *Nymphoides* near Streatham, feels again the thrill that Ray and Nidd must have felt on the self-same spots nearly 300 years before.

Ray's self-banishment from Cambridge two years after the publication of the Catalogue cut short a promising university career; but it left him free to accomplish far more for science than he might well have done had he remained entangled in academic routine. Especially did this freedom give Ray a chance to continue and extend the botanical rides through Britain which he had begun in 1658, and so to build up a first-hand knowledge of our flora incomparably greater than that of any of his predecessors—greater indeed, than that of all but a few of his successors. On these journeys through Britain, and on others through Europe, Ray was often accompanied by his great friend Francis Willughby, a rich young amateur naturalist who helped to support Ray in his new-found poverty and who, on his death in 1672, left Ray an annuity of £70.

There is no space here to describe, even in outline, Ray's half-dozen or so British tours, which included visits to Scotland, Wales, Bristol and Cornwall, and which culminated in the publication of his *Catalogus Plantarum Angliae* in 1670. A reading of his own detailed diaries (Lankester, 1846) helps us to picture him and Willughby riding their twenty to thirty miles a day side by side over the rough seventeenth-century tracks, resting by night at inns or an occasional country-house, absorbing the architectural beauties, local customs and dialects of the country through which they passed, and, above all, peering ever to right and left for the unfamiliar plant that would tempt them from the saddle on to their knees by the roadside.

In Ray's time there were, of course, many now well-known British plants still undiscovered. A fortnight's visit to Yorkshire and Westmorland, for example, yielded first records of *Saxifraga oppositifolia* (on the summit of Ingleborough), *Bartsia alpina* (near Orton) and *Galium boreale* (near Orton and at Windermere); those, indeed, were the days!

The *Catalogus . . . Angliae*, of which a second edition was published in 1677, did for Britain as a whole what Ray's earlier book had done for Cambridgeshire. To-day we see it as a preliminary sketch for his main work on British plants—the *Synopsis Stirpium Britannicarum* (1690) —the first and long-awaited complete descriptive British Flora. Into this famous book, which remained without rival until Hudson's *Flora Anglica* of 1762, Ray poured the accumulated results of his own tours, the reports of correspondents, and the published work of his predecessors, illuminating the whole with the clear light of his genius.

Ray's published work on the British flora closed with some county lists contributed to Camden's *Britannia* in 1695 and a second edition of the *Synopsis* in 1696; he died in 1704. His career, therefore, rounds off the history of British plants up to the end of the seventeenth century. Clarke (1900) estimates that Ray himself added exactly 200 species to our flora, which, together with about half a dozen discoveries by his contemporaries, bring the total at that date up to about 970 species.

It has been said that, in the early days of the exploration of a country, botanical geography is a record of the distribution of botanists rather than of plants. An examination of the early records of British plants well illustrates this point. No less than 60 per cent of the " first records " published before 1700 are from London, Kent or Cambridgeshire. This, of course, reflects the work of Turner, Gerard, L'Obel, Johnson, Merret and others in the London area, of Johnson in Kent, and of Ray in Cambridge. Other areas well represented are Yorkshire (the Rev. Walter Stonehouse and others), Northumberland (Turner), Somerset (Turner and others), Hampshire (Goodyer) and north Wales (Johnson, Ray and others).

Although Ray dominated British field botany at the end of the seventeenth century, he did not, of course, stand alone. Robert Morison (1620-1683), the first professor of botany at Oxford and Ray's rival for botanical pre-eminence at this time, was concerned with the classification of plants as a whole rather than with the study of the British flora; but there was a group of field botanists, largely inspired by Ray's example, that did much good work in the London area, bequeathing to posterity valuable manuscript records and dried specimens, now in the Natural History Museum at South Kensington. I have not space to do more than mention their names—Leonard Plukenet (1641-1706), Samuel Doody (1656-1706), James Petiver (c. 1658-1718) and Adam Buddle (c. 1660-1715)—and to draw attention to a fascinating account of them in Trimen and Dyer's *Flora of Middlesex* (1869).

The period from Ray's death in 1704 until the introduction of the Linnean system into Britain about 1760 was a lean one for the study of British plants. It is not fanciful, I think, to trace a connection between this check and the prevailing temper of British—and indeed European—culture during the same period. The lack of spontaneous appreciation of nature among writers at this time is a commonplace of literary criticism. The proper study of mankind was man. Ray's passionate devotion to " the rich array of springtime meadows " had become, sixty years later, an unforgivable impropriety. Never, perhaps, has the standard of aristocratic culture been so high, but it was a classical and humanistic culture, unfriendly to the study of nature. Nevertheless, some progress was made, though, for the most part, as before, by professional botanists or medical men. It was not until the heyday of the Romantic Revival in the early years of the nineteenth century that the great army of amateurs was to set out on its irresistible forward march.

One of the chief centres of activity in this " post-Ray " period was the Society of Apothecaries' Garden on Chelsea Embankment, which was founded in 1673 and still flourishes there. In 1722 Sir Hans Sloane had put the garden on a sound financial footing by purchasing the freehold from Lord Cheyne, and a little later, under the management of Isaac Rand, and the great Philip Miller, author of the *Gardener's Dictionary* (1731), botanical excursions, or " herborisings " as they were called, were organised in the neighbourhood of London. These acted as a stimulus and an example, not only in this country, but also to botanic gardens on the Continent, and were carried on for many years.

But the two leading botanical figures of the period were Jacob Dillenius at Oxford and John Martyn at Cambridge; and both concerned themselves to some extent with British plants as well as with wider botanical work. Dillenius was born in Germany in 1687 and was brought over to England in 1721 by William Sherard, a friend of Ray and, for a time, consul at Smyrna. Sherard, himself no mean botanist, died in 1728 and in his will left £3,000 to endow the Chair of Botany at Oxford, stipulating that it should be re-named the Sherardian Professorship (as it is to this day) and that Dillenius should occupy it at the first opportunity. This occurred in 1734 and Dillenius remained in the Chair until his death in 1747. His chief botanical labours centred round the mosses and other cryptogams (he published his monumental *Historia Muscorum* in 1741) and on a continuation of Bauhin's *Pinax* (a list of all the then known plants) which still lies in manuscript at Oxford; it is not surprising, therefore, that he had little time and energy to spare for the British flora. On his arrival in England, however, he evidently determined to serve at once his adopted country and set about producing a new (third) edition of Ray's *Synopsis*. For this, he drew on the work of several botanists who had been active since Ray's second edition (1696)—notably Richard Richardson in Yorkshire and elsewhere, Edward Lhuyd in Wales, Samuel Dale in Essex, and Isaac Rand at Chelsea. The work was published in 1724 and remained the only important British Flora until Hudson's *Flora Anglica* (1762). In 1726 Dillenius toured North Wales with his friends Brewer and Brown, but after this he seems to have become absorbed in his wider botanical interests.

John Martyn provides a link between the various rather scattered activities that made up British field botany during the first half of the eighteenth century. He was born in London in 1699 and, though forced into business, botany was his real love. He became friendly with Sherard and Dillenius, studied at Chelsea under Isaac Rand, and was one of those who were stimulated by the London " herborisings " to explore farther afield; by 1730 he had covered many hundreds of miles on foot in southern England and Wales, collecting, observing and recording. In 1733 he was elected professor of botany at Cambridge, where he had already been lecturing for some years. He retained the Chair until 1761, but botany at Cambridge—as elsewhere—was at a low ebb; after 1734 he delivered no lectures, as not a single pupil wished to hear him. Besides an important general history of plants and other works, Martyn published two books on the British flora, an English adaptation (1732) of Tournefort's *History of*

Plants near Paris and a Cambridge Flora (1727). His achievements in field botany were admittedly not great, but he was one of the few to keep the torch flickering between Ray and the Linnean revival in the 1760's. Three others are also remembered (chiefly by book-collectors) for the Floras—now rare—that they published: William Blackstone (d. 1753), author of an account of the plants round Harefield (1737) and of *Specimen Botanicum* (1746—based on Ray's *Synopsis*); George Deering (? 1695-1749), a Nottingham doctor who wrote an entertaining catalogue of plants in the neighbourhood of that city (1738); and John Wilson (d. 1751), who published an English edition of Ray's *Synopsis* in 1744.

By the middle of the eighteenth century Ray had been dead for fifty years, and no fresh breeze had arisen to blow away the cobwebs that were fast stifling British field botany. No fresh breeze, that is, in Britain itself, but across the North Sea, in Sweden, Linnaeus was building his system of classification that was soon to take the botanical world by storm. The details of this system and its influence on taxonomy belong to the general history of botany. Here it must suffice to say that it replaced Ray's tentative and complicated attempts at a natural classification by a simple—but highly artificial—system based on the number of styles and stamens in the flower. Despite this attractive simplicity, it is doubtful whether the system would have made such universal appeal had it not been coupled with Linnaeus' substitution of the "binomial" (or "biverbal") method of naming plants for the cumbersome "polynomial" (or "multiverbal") method in use before him—and with his masterly power of discriminating and describing individual genera and species. It is not surprising that botanists should have grasped thankfully at the opportunity of calling the lesser periwinkle plain *Vinca minor*, rather than "*Clematis Daphnoides minor, seu Vinca Pervinca minor*"—especially when, in the same work, an elegant description separated it neatly and authoritatively from the greater periwinkle, *Vinca major*.

No specific year marks the introduction of the Linnean system into Britain. It began to be widely known in the 1750's through the translation of a number of Linnaeus' books, and in 1760 the self-styled "Sir" John Hill re-arranged Ray's *Synopsis* according to the Linnean system. This was a poor book in many ways, and the first real bridgehead established by the new system was the publication of William Hudson's *Flora Anglica* in 1762. This at once superseded

Ray's *Synopsis* and remained the chief British Flora for many years. Ray and his natural system were eclipsed and Linnaeus reigned supreme.

What was the effect of this revolution on field botany in Britain? Briefly, it put into the hands of the many an easily used botanical method to replace the clumsy and forbidding pre-Linnean apparatus which all but the few had shrunk from using. " Herborising " became popular, even among what was then known as the " fair sex." A delicious exposition of field botany as an elegant female accomplishment can be found in the preface to Abbot's *Flora Bedfordiensis* (1798). " That such excellence," he writes, " should have been attained in this branch of science by so many of the female sex, notwithstanding the disadvantages they labour under from the want of scholastic and technical instruction, is a convincing proof of the liberality with which Nature has endowed the female mind: and how little reason there is to suppose that their intellectual are from any other cause than want of cultivation, in any degree inferior to their personal accomplishments."

The result of this fresh wave of activity is well brought out if we compare the number of local Floras that appeared before and after the dividing date of 1760. During the first sixty years of the eighteenth century eight books of this type were published, while during the succeeding sixty years, up to 1820, there were nearly forty.

Many botanists, amateur and professional, played their part in this revival, but it will be best, perhaps, to sketch the careers of two, as samples of the spirit and working methods of the period, rather than to give a bare catalogue of many names, dates and books.

William Curtis (1746-1799) is remembered to-day as the founder of the *Botanical Magazine* in 1787 ; this noble periodical, with its lovely coloured plates of garden flowers, has appeared continuously ever since. But, in addition to his work for cultivated plants, Curtis was one of the leaders of field botany in the early Linnean period. He was born at Alton in Hampshire, the son of a Quaker tanner, and early broke away from training as an apothecary to follow his passion for botany, becoming *Praefectus Horti* at the Chelsea Physic Garden in 1772. Already at Alton a botanically-minded ostler at the Crown Inn had inspired him with a love of wild flowers and together they had explored the Hampshire countryside. He continued his herborising around London and soon conceived the idea of his great work, the

Plate I. N. D. Simpson, H. W. Pugsley and A. J. Wilmott examining Salicornias on a Botanical Society excursion to Sheppey, 1938

Plate II. The sedge *Carex pendula* in woodland glade, Charnwood Forest, Leicestershire, June

Flora Londinensis, which appeared in folio parts from 1775-1798. Its chief appeal, then as now, was the series of magnificent coloured plates, 435 in all, drawn and engraved by Sydenham Edwards, James Sowerby and other famous artists, and hand-coloured by a team of 30 colourists. But the text is of equal or greater value. Curtis was an acute and thorough botanist, bringing order out of chaos in many difficult groups; Sir James Smith ranked the *Flora Londinensis* as " next to Ray's *Synopsis* in original merit and authority on British Plants."

Curtis left Chelsea in 1777 to concentrate on running his own botanic garden, which he moved from Bermondsey to Lambeth and later to what is now the Fulham Road. Apart from the *Flora Londinensis* he did not publish much on British plants, but, until his death in 1799, he remained, through his friendships and correspondence, at the centre of the ever-increasing botanical activity of this period.

Sir James Edward Smith (1759-1828), whose praise of Curtis I have quoted, was born in surroundings very different from the Alton tannery. His father was a rich Norwich manufacturer and, though trained for medicine, James could afford to devote his working hours to botany. It was, in fact, his father's thousand guineas which launched him with considerable aplomb on the botanical world—as the purchaser, at the age of 24, of Linnaeus' collection of plants and library. These had been offered to Sir Joseph Banks, President of the Royal Society, and it was at breakfast with the great man that Smith decided to buy. A few years later, in 1788, at a Coffee House in Great Marlborough Street, the Linnean Society was founded by Smith and other botanists, and the collections and books are still housed in the Society's Rooms at Burlington House—a lode-star for naturalists from all over the world. In the Society was concentrated the botanical (and zoological) activity of the time, and it was not until towards the middle of the nineteenth century that other societies and institutions began to share the burden.

Smith, unlike Curtis, was a voluminous writer on the British flora. His best remembered and most influential book is rather unfairly known as " Sowerby's *English Botany*," published in 36 volumes between 1790 and 1814. James Sowerby drew the beautiful and accurate plates (which make a good set worth a great deal to-day), but Smith wrote the text—the first large-scale account of our flora and the foundation of all later work. Smith was President of the Linnean Society for forty years, until his death in 1828; he thus spanned the heyday, and lived to hear the death-knell, of the Linnean ascendancy in Britain.

However much the artificiality of Linnaeus' system may have stifled the growth of a natural, scientific classification of plants, its triumph undoubtedly transformed British field botany. By 1820 the meagre tale of workers of the 1750's, struggling with Ray's " polynomial " *Synopsis*, had been replaced by groups of amateur plant-hunters all over the country, equipped, thanks to Smith and other writers, with a number of easily consulted books, adequate for identification. But these books—and the local Floras of the time—were arid, rather forbidding catalogues, often written in Latin, hardly hinting at the wave of popular literature that in the middle years of the nineteenth century was to carry field botany into half the country vicarages and consulting rooms in the land. Linnaeus, with his system and his simple nomenclature, made botany popular in spite of the classics-bred inertia of his time, but it needed the wider and deeper influence of the Romantic Revival at the turn of the century to clothe the bare bones of his orders, classes, genera and species with the magic garment of nature-worship.

The artistic aspects of this Revival—the poetry of Wordsworth, the music of Schumann, the painting of Delacroix—have been many times examined and expounded. Undoubtedly, however, it contained also a powerful " back-to-nature " element, and it was this that caused the phenomenal expansion of field botany about 1820. To cull the flowers in Ray's " rich array of springtime meadows " had once more become respectable. We may again take the production of local Floras as a sensitive index of this expansion. We have seen that from 1760 to 1820 about 40 were published; from 1820 to 1880 there were 170!

The change of heart in the 1820's was accompanied by the death of Linnaeus' system (though not of his nomenclature), and the rebirth of a natural classification based on the system of Ray that had lain dormant for fifty years. The reaction against Linnaeus began in France towards the end of the eighteenth century, but many hands in many countries were to shape (and are still shaping) the various natural systems that emerged. In Britain, Bentham and Hooker evolved the arrangement familiar in their popular *Handbook*, and, later, Charles Darwin gave natural classification a new meaning through his theory of evolution.

But the nature-worship of the Romantic Revival was not the sole cause of the nineteenth century expansion of field botany. At least three others accompanied and reinforced it: the re-opening of the Continent after the Napoleonic Wars; the invention of the steam-driven railway engine; and the spread of popular education, as the liberal ideas, kept alive by the opposition Whigs and Dissenters during the reactionary years following the French Revolution, gradually bore fruit from about 1822 onwards. Renewed intercourse across the Channel enabled British and Continental botanists to compare and discuss their respective floras; improved transport carried naturalists to every corner of the country; and the steady increase in the educated population brought more and more workers into the field. The second quarter of the nineteenth century saw the horse-borne pioneer, with a Latin treatise in his saddle-bag, capitulate before the growing bands of top-hatted Victorians entraining for a week's excursion to the Lizard or Teesdale, armed with the latest Flora of Lindley, Hooker or Babington.

The tide of this expansion flowed along three distinct channels. More Floras were written, both national and local, and a spate of popular books appeared on wild flowers, some of them drenched in a sentimentality exhibiting the Romantic Revival at its worst; natural history societies and field clubs sprang up all over the country, pub-lishing journals and local Floras, exchanging specimens, and organising botanising expeditions; and an ever-increasing number of serious and critical individual students of our flora—the majority of them amateurs—set themselves, by their patient and accurate work, to make Britain the best-botanised country in the world.

To trace in detail the history of field botany in the nineteenth century would fill a whole volume—a volume that one day will, I hope, be written; the evolution of local Floras alone would cover many pages. I will try, however, to pick out a few of the landmarks along the three channels I have mentioned, inevitably omitting, I fear, many outstanding names.

Hudson's " Linnean " *Flora Anglica* (1762) was still the standard handbook to our flora when, in the 1820's, the Natural System began to gain ground in Britain. John Lindley, its leading prophet, produced a *Synopsis of the British Flora* in 1829, and Sir William Hooker, the first Director of Kew Gardens, followed with his *British Flora* (1830); but the three books which have proved themselves, in their various editions and modifications, the prop and stay of field botanists until the appearance of Clapham, Tutin and Warburg's *Flora* (1952) were Babington's *Manual* (1843), Bentham's *Handbook* (1858) and Sir Joseph Hooker's *Students' Flora* (1870). Charles Babington, Professor at Cambridge from 1861 to 1895, was the first to treat British plants critically in relation to the flora of the Continent. He was a " splitter " in contrast with George Bentham, who was a " lumper "; that is to say, Babington favoured making many species based on small differences, while Bentham preferred to " lump " several of Babington's species into one " large " one—two methods of approach which, being based mainly on temperament, are likely to persist as long as plant classification. Sir Joseph Hooker, son of Sir William and his successor at Kew, edited the *Handbook* after Bentham's death, thus creating the much-loved " Bentham & Hooker "; Sir Joseph's own *Students' Flora* was perhaps the most scholarly of the three. On a larger scale was Dr. J. T. Boswell Syme's 12-volume edition of *English Botany* (1863-1892) which was an oracle to be consulted in public libraries.

Side by side with Floras of the whole country, local Floras of counties or smaller districts appeared in increasing numbers. These showed, in their style and contents, the growing influence of the broader, more popular approach to field botany that was in the air. The typical eighteenth-century local Flora, for example *Plantae Woodfordienses* (1771) by Richard Warner, was little more than a brief catalogue of names and localities, sometimes entirely in Latin, though Warner makes a concession, as he says in his Preface, " by specifying the soil, the place of Growth, and the time of Flowering, in the *English* language." By the end of the nineteenth century a transformation had taken place, and sumptuous, often illustrated, volumes were appearing with historical introductions, sections on geology and climate, and a full treatment of each species, giving its distribution in detail and information on folk-lore and economic uses. J. W. White's *The Bristol Flora* (1912) is, I think, one of the best specimens of the type, and such fine volumes as the *Flora of Gloucestershire* (Riddelsdell et al. 1948) and the *Flora of Bedfordshire* (Dony, 1953) are carrying on the tradition. The earliest authors to make a marked break-away from the eighteenth-century " catalogue " were J. P. Jones and J. F. Kingston in their *Flora Devoniensis* (1829), but the pattern of the nineteenth-century local Flora became finally stabilised with the *Flora of Middlesex* (1869) by Henry Trimen and William (later Sir William) Thiselton-Dyer, which set a high standard of scholarship and completeness for all later writers. These local Floras, now covering nearly the whole country, are fascinating volumes, reflecting in their pages the history of British field botany from 1629, when Thomas Johnson published his *Iter ... in Agrum Cantianum*, down to the present day. To a botanist infected with the virus of book-collecting they are apt to prove irresistible.

Every movement has its camp-followers, aping the true leaders, but achieving only a caricature of their greatness. Hanging on to the coat-tails of the main army of Victorian field botanists trooped, as I have said, a multitude of writers of popular books on wild flowers in which reverence for nature had degenerated into a ludicrous and pretentious sentimentality. Perhaps the lowest depths were plumbed by Edwin Lees in *The Botanical Looker-out* (1842: 2nd ed. 1851), although, curiously enough, he was responsible also for quite a respectable *Botany of the Malvern Hills*. Here is a sample of his prose: " Whether presenting a bouquet of flowers with courteous smile to his lady-love, or moralising as he hangs over the topmost turret of some princely ruin, to pluck a sweet gem that smiles amidst the desolation there, like an iridean tinge upon a dark cloud, the Botanical Looker-Out is ever at home." A century that has given birth to the pop song can, I admit, ill afford to sneer at the literary curiosities of earlier days, but we can at any rate congratulate ourselves that the botanical looker-out no longer frequents princely ruins nor moralises as he hangs from their topmost turrets.

It was natural, remembering the Englishman's fondness for organising himself into committees, that the increasing popularity of field botany in the nineteenth century should lead to the formation of a spate of natural history societies and field clubs, both local and national. Lysaght (1959) compiled an excellent directory of these bodies, giving, among other information, the dates of their foundation. If we analyse these dates we find that roughly seventy per cent of the societies were founded during the nineteenth century, twenty per cent in the first half and fifty per cent between 1851 and 1900. This mass birth during the second half of the century was a result of the gradual building up of interest in natural history during the preceding fifty years. Incidentally, there seems to have been a revival in the formation of these clubs and societies in recent years; Lysaght's directory shows about twice as many foundations between 1926 and 1950 as compared with the period 1901-1925.

Great as has been the influence of such organised effort, the individual worker of genius was of equal importance, and British field botany during the nineteenth century has fortunately been rich in outstanding personalities. I will select as examples H. C. Watson and G. C. Druce, both amateurs, whose active careers spanned, between them, a period of a hundred years, from 1832 to 1932.

Hewett Cottrell Watson (1804-1881), in the outward circumstances of his life, was a staid Victorian bachelor, residing for nearly fifty years at Thames Ditton in Surrey; but in his botanical writings and controversies he was something of an *enfant terrible*. " Nothing short of a full-scale biography," writes R. D. Meikle (1949), " can hope to do justice to such a rare personality." His outstanding contributions to British botany were in the realm of plant distribution. Comparatively little accurate and systematised information was available on this subject when he published his *Outline of the Geographical Distribution of British Plants* in, appropriately enough, 1832, the year of other and more famous reforms. He followed this with further works in the same field, notably *Cybele Britannica* (1847-59) and *Topographical Botany* (1873-4), in which he divided Britain into the 112 " vice-counties " that are still used, little modified, by botanists to-day. Each vice-county comprises either a whole county (e.g. Cambridgeshire) or, in the case of the larger counties, a part of a county (e.g. Sussex is divided into two vice-counties, East Sussex and West Sussex). These divisions provided a generally accepted and stable method of recording the distribution of British plants and proved a great stimulus to a more accurate study of botanical topography in Britain. Watson's main interest was British botany, but he entered many other fields with equal combative gusto, and the journals of the time bear witness to his vigorous mind in such diverse subjects as phrenology, Darwinism, and radical politics. At the end of his long life he regretted his earlier fireworks. " They read too bitter," he remarked; " you don't know what it is to have a large organ of destructiveness."

Watson stands at the centre of the great mid-nineteenth-century development of British field botany—a development concerned primarily with the nomenclature, classification and detailed distribution of British plants. George Claridge Druce (1850-1932) continued this tradition into the fourth decade of the twentieth century and can, perhaps, be looked upon as its last great figure, standing somewhat aside from the new trends in field botany which, as we shall see, had their birth about the beginning of the century.

Druce's career was a remarkable one. Beginning as a poor boy with little education, he became an M.A. of Oxford at thirty-six, Mayor of the City at fifty and was elected a Fellow of the Royal Society a few years before he died. For over fifty years he was proprietor of the pharmaceutical chemist's business in the High Street which, until recently, still bore his name.

Druce's botanical activities were legion, but he will perhaps be best remembered for his long secretaryship of the Botanical Society and Exchange Club of the British Isles (now The Botanical Society of the British Isles), for his *Floras of Oxfordshire, Berkshire, Buckinghamshire* and *Northamptonshire*, for his editions of Hayward's *Botanist's Pocket Book*, for his strong views (often in opposition to the majority) on plant nomenclature, and for his naming of many minor variations of British plants—a policy which, while certainly putting such variations on record, was frequently criticised as an unnecessary multiplication of names. He built up a magnificent Herbarium and Library now housed at the Department of Botany at Oxford. One of Druce's most striking characteristics was his eager and generous help to young botanists. I still treasure an enthusiastic and nearly illegible letter that I received from him, by return of post, when as a schoolboy of fifteen, I had sent him a specimen of *Galium anglicum* from Northamptonshire—a new county record; and many other similar letters, equally treasured and equally illegible, must be in existence. His death in 1932 closed an era in British field botany which had opened nearly a hundred and fifty years earlier with the great expansion of activity inspired by the Romantic Revival.

At the beginning of the twentieth century, as I have said, developments occurred which were to open up new vistas in the study of plants, and to-day, as a result, there are two fairly distinct but interacting lines of work being carried out on the British flora. The first is a continuation of what might be called the Watson-Druce tradition and is admirably illustrated by the remarkably successful Maps Scheme; while the second, deriving from the establishment of the sciences of ecology, genetics and cytology fifty years ago, has, again, two distinct but interacting aspects, which may be broadly distinguished as ecology " proper " and experimental taxonomy. A brief description of the meaning and content of these is given in Chapter 14. The modern development of ecology, or the study of plants in relation to their environment and of the structure of plant communities, can be dated from the publication of Warming's *Plant Communities* at Copenhagen in 1895 and the new science was soon taken up enthusiastically in Britain. Prof. Sir Arthur Tansley published his *Types of British Vegetation* in 1911, and in 1939 the results of many years' ecological work by himself and others was set out in his monumental volume, *The British Islands and their Vegetation.* Much work in descriptive ecology is still being done in Britain, and The British Ecological Society, founded in 1913, provides a focal point for such work. The relationship between ecology and systematic botany is, of course, a very close one, since it is essential for the ecologist to know accurately the species and varieties making up his communities and, conversely, a knowledge of ecology is of great help to the systematic botanist in the understanding of the distribution of the plants with which he is working.

Experimental taxonomy is a product of the combined influences of ecology, genetics, cytology and systematic botany, against an evolutionary background deriving from the acceptance of Darwinism nearly a hundred years ago. In contrast with the older Watson-Druce tradition of descriptive systematic botany, it looks at plants from a more dynamic point of view, seeking to discover how the various units have arisen in the immediate past and to establish the evolutionary relationship between them. Much detailed information on methods and results in experimental taxonomy will be found in Dr. W. B. Turrill's volume in this series (1948). Historically, the discovery, at the beginning of this century, of Mendel's work that had been carried out fifty years earlier and since forgotten, was the starting point for the rapid development of genetics and cytology during the following thirty years, and experimental taxonomy, as I have said, grew from a combination of these with ecology and systematic botany. One of the first and most influential pieces of work was Turesson's experimental investigation in Sweden of the ecological forms within a number of Scandinavian species. Those forms which he found to be genetically distinct, he called " ecotypes " (see p. 44) and his establishment of the principle of ecotypical differentiation inspired considerable work along the same lines in this country, notably the transplant experiments carried out by Marsden-Jones and Turrill at Potterne, and Gregor's researches on *Plantago maritima* and other species at Corstorphine. Since this pioneer work, an ever-increasing number of British botanists has been attracted to experimental taxonomy, especially at the universities, where systematic botany has received a welcome fillip since the experimental method came to the fore. The *Biological Flora*, now being published in parts by the Ecological Society, will form an invaluable repository for this new type of knowledge of our flora, and many papers are appearing in botanical journals on individual species and genera.

I do not wish to over-emphasise the differences, marked as they are, between the experimental and the older type of taxonomic work, nor in any way to minimise the importance of the latter. The charting of the morphological units in any flora is a basic necessity before experimental—or any other—investigations can be undertaken, and there is still a great deal of work to be done along the traditional lines, especially on the detailed distribution of even common species and in the correlation of British with Continental forms (see Chap. 14). The fact that three distinct and well-marked species have been discovered, or re-discovered, in Scotland during recent years (see p. 124) speaks for itself. There is no doubt that both currents—the old and the new—in the broad stream of British field botany will, for many years to come, contribute to the forward flow that was started 400 years ago by William Turner when he first saw his bluebells " muche aboute Sion." The study of British plants has never been more flourishing, nor more full of promise, than we find it in these middle years of the twentieth century.

CHAPTER 3

THE BIOLOGY OF OUR FLORA

(J. G. *and* M. W.)

THE WORD " BIOLOGY " has unfortunately acquired a double meaning. It is used most frequently to denote the science of living things (i.e. botany plus zoology), but it has also come, in such phrases as the " biology of flowering plants," to mean all the activities and functions of an organism connected with its passage through life, from the cradle to the grave. That is to say it includes, in a flowering plant, among other things, germination and subsequent development to adult size, the production of flowers, pollination, fruit and seed production and dispersal, vegetative propagation, methods of over-wintering, and span of life. The second meaning may also be extended to include, though this usage is not fixed, the evolutionary genetics and cytology of a species, and its relation to the environment, including soil, climate and the other organisms with which it grows (i.e. its ecology).

It is in this last and broadest sense that we have used the word biology in the title of the present chapter. It is clearly impossible in the space available to cover such a wide field in more than a very brief and summarised fashion, but, within these limits, we have tried to build a sort of " biological framework " into which the facts about the individual species dealt with later can be fitted.* A much fuller treatment will be found in Turrill's volume in this series, *British Plant Life* (1948). The chapter is divided into three sections: (i) life-histories, (ii) the individual, the community and the environment, and (iii) species and evolution.

*We have omitted all reference to basic plant physiology and morphology, which are dealt with in general text-books of botany.

1. Life-Histories (J. G.)

Let us start with the seed in the ground. This does not imply that the seed represents the beginning of a new individual; it is, of course, merely a stage in the individual's history, a history which began with the single-celled fertilised ovum or egg inside the ovule of the parent plant (see p. 34). This ovum has now developed into a tiny embryo shut up inside the hard covering of the seed, and on the safe custody and eventual growth of this embryo the future existence of the individual depends. The seed, therefore, is one of the most crucial stages in the life of a plant, and the devices which enable it to perform its to some extent antagonistic functions of portable strong-box and wet nurse for the embryo are complicated and fascinating. It must persist unharmed through many chances and changes unfavourable or harmful to the growth and development of the embryo, and yet make a successful response as soon as favourable conditions occur. This peculiar defensive-offensive role has resulted in an exceedingly complex set of reactions by the seed to the various factors of the environment, such as humidity, temperature, light, and aeration.

The chief defensive weapons of a seed are exterior hardness and interior dryness (round about 10 per cent of water as against up to 40 per cent in a germinating seed). This combination enables it to withstand extreme temperatures ($-234°$ C. and $100°$ C. have been tried on some seeds with impunity), to survive prolonged periods of drought, and to resist attacks by fungi. The length of time which different seeds can remain in this defensive state of suspended animation varies enormously, from the few days of willow seeds, through the fifteen or so years of wheat (despite the stories of the germination of " mummy " wheat from Egyptian tombs), to the perhaps several hundred years of the lotus (*Nelumbo nucifera*) seeds brought from Manchurian peat in 1951 and successfully germinated in America (Youngman, 1951). As a secondary defensive mechanism, seeds find safety in numbers and can well afford to lose a large proportion of the often enormous production of a single plant (see p. 32).

Sooner or later, if it survives, a seed goes over from the defensive to the offensive and germination takes place. The timing of this change depends on both external and internal factors. In some plants, e.g. *Crataegus* spp. (hawthorn), the embryo is not sufficiently developed, when the seed is shed, for germination to take place, even if external conditions are favourable; in others (e.g. sweet peas), though the embryo is ready, the covering of the seed is so resistant to water that some considerable time elapses before enough can be absorbed for germination, while in others (e.g. willows) germination occurs almost immediately the environment is suitable. What is a suitable environment? The answer varies from species to species, but the prime factors are (1) enough humidity for water to be absorbed through the seed coat (too much water, however, will prevent germination through lack of oxygen), and (2) a temperature ranging between certain limits, the total range varying from just over freezing point to over 40° C.

A secondary, but nevertheless important, survival device is the fact that, in many species, not all the seeds shed by one individual germinate at the same time: various types of this delayed germination occur and link up with the span of life of the individuals concerned and with the habitat and distribution of the species. The easiest way to understand this link-up is to begin with the two broad divisions based on span of life, namely monocarpic and polycarpic plants. A monocarpic plant is one that produces flowers and fruit only once in its lifetime and then dies, whereas the individuals of polycarpic species flower and fruit more than once, some of them—for example, trees—annually over a period of many hundreds of years. Monocarpic species may be summer-annuals (germinating in the spring and flowering the same year, e.g. the speedwell *Veronica hederifolia*), winter-annuals (germinating in the autumn, over-wintering as rosettes, and flowering the following year, e.g. the buttercup *Ranunculus parviflorus*), biennials (germinating in the spring, forming a rosette during the summer and flowering the following year, e.g. the mullein, *Verbascum thapsus*), or what are often mis-called " century plants " (remaining several years in a vegetative condition before finally flowering, fruiting and dying, e.g. many agaves and bamboos). A few summer-annuals (for example, shepherd's purse, *Capsella bursa-pastoris*), may produce two or even three generations in a single year (ephemerals).

The distinction between summer and winter annuals is not, however, clear-cut. Some species can function as both, and even as biennials as well, and it is here that germinating habits link up with span of life. Three types of germination occur. All the season's seed of a species may germinate at once (simultaneous germination); in another species, there may be two or more bursts of germination, separated by considerable intervals (discontinuous germination); and lastly, germination may be more or less uninterrupted over a long period, whenever conditions are favourable (continuous germination). Now, a species like *Ranunculus parviflorus*, with simultaneous germination in the autumn, is clearly a " pure " winter annual, while *Veronica hederifolia*, with simultaneous germination in the spring, is a " pure " summer annual. The corn gromwell (*Lithospermum arvense*), on the other hand, is a continuous germinator throughout the year, and in such a species, some individuals must obviously behave as summer annuals and some as winter annuals. Again, in the centaury (*Centaurium erythraea*), which has two bursts of germination, one in the autumn and another in the spring, there are both winter and summer annual individuals and, in addition, certain plants (germinating in the spring, but not flowering until the following year) behaving as biennials.

This interrelation between germination and span of life is a fascinating subject and one that links up closely with the capacity of a species to survive under adverse conditions and hence with its distribution. Sir Edward Salisbury, for example, records that, in his garden at Radlett, during the severe frosts in the early months of 1929, all the seedlings of *Ranunculus parviflorus* and the little rock-rose *Helianthemum guttatum* subsp. *breweri*, which had germinated the previous autumn, were killed. The *Ranunculus*, being a simultaneous germinator, had nothing left up its sleeve and was exterminated, whereas the *Helianthemum*, which germinates in three distinct batches, had one more batch to come after the frost and survived quite happily. There is much more to be learned about the germination and span of life of many quite common British species, and here is a first-rate field of work for amateur botanists with patience and a gift for experimentation.

Apart from these differences in times of germination, the proportion of the total seed produced that finally germinates at all, even under perfect conditions, varies considerably from species to species, as gardeners know to their cost. Tests of the three British species of sundew, for example (Salisbury, 1942), have shown that *Drosera rotundifolia* has a germination of about 30 per cent, *D. anglica* about 18 per cent, and *D. intermedia* as low as 2 per cent. In *Silene conica*, on the other hand, as many as 98 per cent of the seeds normally come up. Here again, there is a great need for further experiment, not difficult to carry out, especially by those with a greenhouse.

When a seed has germinated its troubles are by no means over. Very little is known about the exact mortality among seedlings growing in the wild, but several causes combine to make it high. Tests have shown that, even under laboratory conditions, a number do not survive; in one experiment 6 per cent of *Silene conica* seedlings died. In nature, many must succumb to disease, as they do in seed-boxes; but the greatest slaughter is due to competition with fellow-seedlings for the limited food, water and light available, resulting in the death of many plants vigorous and healthy enough to have survived if growing alone. That this drastic thinning does actually take place is obvious when we consider that a single plant of mullein (*Verbascum thapsus*), which has an 80 per cent germination, may produce as many as 160,000 ripe seeds; if the whole 80 per cent grew into adult plants, they would cover an area the size of a football pitch. In a close carpet of seedlings the struggle for survival is at its keenest.

Some seedlings, however, do survive this struggle and grow up to produce leaves, flowers, fruit, and a further crop of seed. In all flowering plants, during the early period of their growth, only stems and leaves are formed, but sooner or later the minute beginnings of flower-buds are laid down. This swing-over to flower formation depends on a number of factors, both inside and outside the plant. There must not only be sufficient light and heat, but in many plants the duration of light during the 24-hour period must also be just right, or they will not flower. There are " short-day " and " long-day " plants. The former, which are mostly natives of low latitudes where the daily periods of light and darkness are approximately equal, will not flower while there are more than 12 hours daylight (e.g. *Cosmos*, from Mexico), whereas the latter, from higher latitudes, behave in the opposite way, and will flower only if the period of daylight exceeds 12 hours (e.g. *Rudbeckia*, from N. America). Some plants, on the other hand, are indifferent to this factor of day-length.

The relative amounts of nitrogen and carbohydrates in a plant also influence the production of flower buds. Broadly speaking, excessive nitrogen favours continued vegetative growth, while an excess of carbohydrates stimulates flowering. In trying to make sense of the factors that influence flower-production we must remember that in many plants flower buds are formed several months before the flowers actually appear. Hazel catkins are laid down during August or September of the previous year, and it is the conditions at that time (for instance the nitrogen/carbohydrate ratio) that affect the abundance or otherwise of flowers the following February.

Some years ago Professor Went in Holland discovered that there are substances in plants corresponding more or less to the hormones of animals—substances which travel about the body controlling some of the vital processes. These plant hormones affect growth in many ways, including the formation of flowers and fruit. They have been developed commercially as stimulators of rooting in cuttings, and, more recently, as selective weed-killers and in other ways; on lawns they induce overgrowth and death in many weeds, while leaving the grass unharmed.

The function of a flower in the life of a plant is the production of seed; its colour, its scent, its elaborate parts and intricate mechanism all serve this end. Seed cannot normally develop, of course, until fertilisation has taken place, and to the task of ensuring that pollen reaches a ripe stigma, that male cells are carried down towards the ovule by the pollen tube, and that a male nucleus eventually fuses with a female nucleus, Nature has brought all her customary lavishness of invention. The transference of pollen from the stamens to the stigma is called *pollination*, and the events that follow, terminating in the fusion of the male and female nuclei, are called *fertilisation*.

Most people have a somewhat hazy knowledge that " fresh blood " is a good thing for the preservation or increase of vigour in living things. " Fresh blood," in flowering plants, means *cross-pollination*, that is, the carrying of pollen from the stamens of one individual to the stigmas of another individual; in *self-pollination* the pollen comes from stamens in the same flower or in a different flower on the same individual. The complicated and various devices that exist to ensure cross-pollination and to prevent self-pollination make sense when we realise that cross-pollination is the means whereby new gene-combinations (see p. 43) are brought together, thus forming the raw material for evolutionary change. That cross-pollination is not, however, essential for short-term survival is shown by the fact that many species are exclusively self-pollinated or reproduce themselves by apomixis (see p. 45).

Pollen grains cannot move by themselves, and some outside force must carry them from the stamen of one flower to the stigma of another:

this force may be provided by wind, insects, water, or, in some species, birds or bats. Some of the different ways this journey is undertaken and the floral mechanisms associated with them are described in later chapters, but I will mention here one broad distinction between flowers pollinated by animals, and those pollinated by non-living agents such as wind or water. The former, in general, possess characteristics such as bright colours, scents, or the secretion of nectar, which are attractive to insects or birds, while the latter lack these attractions, but show other features, for example abundant pollen easily scattered by the wind, which fit them for their particular method of pollination. The large, sweet-scented, yellow flowers of the evening primrose (*Oenothera*), opening at dusk to receive the visits of the night-flying moths which pollinate them, contrast strikingly with the tiny male flowers of the hazel, crowded together into dangling catkins which emit a cloud of pollen grains at the slightest puff of wind.

Although cross-pollination is so elaborately catered for in flowering plants, and in the considerable number of " self-sterile " species (e.g. cuckoo-flower, *Cardamine pratensis*) fertilisation does not take place even if pollen falls on the stigma of the same flower, nevertheless, as I have said, self-pollination and self-fertilisation do occur fairly frequently. In some species which are normally cross-pollinated, self-pollination takes place if cross-pollination fails, while in others, like the oat, barley and wheat, where self-pollination takes place as soon as the flowers open, cross-pollination never, or only very rarely, occurs naturally. The most extreme case of self-pollination is that of " cleistogamous " flowers, where development appears to be arrested at an early stage with the result that the flowers never open properly. No foreign pollen can reach their stigmas and self-fertilisation is thus inevitable. In a number of species, for example the sweet violet (*Viola odorata*), both ordinary " open " flowers and cleistogamous flowers are produced (see p. 91), while others have only cleistogamous flowers (e.g. the sage *Salvia cleistogama*).

By whatever method the pollen may have reached the stigma, whether by cross- or self-pollination, the next event is the growth of a tube from the pollen grain, through the tissues of the stigma and style, into one of the ovules enclosed in the ovary. Down this tube pass two male nuclei, one of which fuses with the egg cell nucleus to produce, by cell division, a new embryo; and the other with two

Plate III. Stinking Hellebore, *Helleborus foetidus*, Hartington Dale, Derbyshire, April

R. H. Hal

Plate IV. Herb Paris, *Paris quadrifolia*, with Dog's Mercury, *Mercurialis perennis*. Via Gellia, Derbyshire, June

special nuclei in the ovule to form the endosperm—the nutritive tissue which provides food for the embryo as it develops inside the seed.

The transport of the ripe seed from the fruit to a site suitable for germination is a problem not unlike the transport of pollen, and it is solved in broadly similar ways—that is, by the aid of animals, wind and water, but also, sometimes, by an explosive mechanism in the fruit itself, as in balsam (*Impatiens* spp.). Here again particular cases are described in later chapters, and some of the devices, though hardly rivalling the ingenuities of pollination, are remarkable enough, for example, the neat parachutes of the composites (see Plate 1 p. 38) drifting lazily in the summer sunshine, and the burrs of goose-grass (*Galium aparine*) clinging so tenaciously to trousers or stockings.

So far we have been dealing chiefly with the life-histories of annuals and biennials which pass the winter as seeds, or as rosettes destined to flower and die the following year. Perennials have evolved many other methods of over-wintering, and the Danish botanist Raunkiaer has devised a classification of flowering plants based mainly on over-wintering methods, coupled with form and place of growth, which is extremely useful, especially in demonstrating how these different *life-forms*, as he called them, link up with the type of environment in which the plants are growing. Raunkiaer's life-form classification is summarised in Clapham, Tutin and Warburg's *Flora*, and under each species its life-form is stated. It may be useful, however, to give here the seven main divisions, most of which are further subdivided in the full scheme.

1. Phanerophytes—woody plants with buds more than 25 cm. above soil level (e.g. shrubs and trees).
2. Chamaephytes—woody or herbaceous plants with buds above the soil surface but below 25 cm. (e.g. dwarf shrubs).
3. Hemicryptophytes—herbs (very rarely woody plants) with buds at soil level (e.g. many herbaceous perennials).
4. Geophytes—herbs with buds below the soil surface (e.g. plants with bulbs or corms).
5. Helophytes—marsh plants (e.g. many sedges and rushes).
6. Hydrophytes—water plants (e.g. pond-weeds and water-lilies).
7. Therophytes—plants which pass the unfavourable season as seeds (e.g. annuals).

2. THE INDIVIDUAL, THE COMMUNITY AND THE ENVIRONMENT
(J. G.)

I have sketched very briefly the life-history of an individual flowering plant from embryo to embryo; but the individual does not live in splendid isolation. It is always a member of a community, and we cannot fully understand its behaviour unless we study its relation to the community and to the environment in which the members of the community exist.

Let us consider, first, the environment, which may be defined as the sum total of the conditions under which a plant community lives. Obviously some of these conditions are more important than others in determining the type of plant that can grow in a particular habitat, and it is convenient to divide these " determining factors " into three groups:

1. *Climatic Factors.*
2. *Edaphic or Soil Factors.*
3. *Biotic Factors*, or those due to the action of living organisms including man.

Climate is clearly the most important influence in determining the broad type of vegetation occurring in any area. Whatever the soil, tropical rain forest could never be made to grow in Hyde Park, nor a birchwood by the shores of the Red Sea. The word " climate " is a convenient shorthand for a number of factors, the most influential of which on plant growth are temperature, light, humidity and wind. These must not, of course, be thought of in isolation from each other. They interact closely; for example, both wind and temperature, in addition to rainfall, have a strong influence on humidity. The actual mechanism of the action of climatic factors in determining which plants shall or shall not grow in any area varies from comparatively simple and direct effects like the killing of young growth by spring frosts, to the much more complex interaction between air humidity and transpiration, or water-loss, from leaves. Although, in broad outline, these mechanisms are fairly well understood, there is still much to be learnt about their detailed method of operation in particular plants.

As I have said, climate determines the main type of vegetation growing in any area. For example, the warm, fairly damp summers and cold winters of Western Europe are favourable to the development of deciduous forests, rather than the broad-leaved evergreen forests of tropical climates, since water-loss through large winter-persistent leaves at a time when little water can be absorbed by the roots from the frozen soil would be very harmful. But, in addition to these wide climatic effects, local variations in the topography may lead to local differences in climate which often have marked effects on vegetation within the main climatic type. For example, the northern and southern slopes of a mountain may carry quite different plants, and on an even smaller scale, the shelter of a cliff or rock may create a " micro-climate " appreciably more favourable to plant growth than that on a more exposed slope a few yards away. In later chapters the influence of climate is considered in more detail in connection with particular habitats in the British Isles.

If climate is the master factor in determining vegetation, *type of soil* is certainly a most important secondary influence in creating local differences within the main vegetational type; one has only to think of a small area like the Isle of Purbeck in Dorset, where chalk and limestone grassland occur side by side with sandy heaths, to realise how striking this influence can be. The actual effect of different soils, like that of climates, is complex and due to a number of factors, including chemical composition (e.g. presence or absence of lime and of salt water), moisture-holding capacity, and the amount of decaying organic matter, or humus, present. Further, the type of soil in any area is determined, not only by the " parental " rock, but also by the climatic conditions under which the soil has been formed. For example, the development of blanket-bog peat (see p. 134) takes place only in cool climates with a high rainfall.

Each of the main kinds of soil in Britain, sands, gravels, peats, silts, clays, loams, etc., has a type of vegetation and individual species which are particularly characteristic of it, and there are many subdivisions of the main types (e.g. calcareous and non-calcareous clays) which, again, bear specially characteristic plants. In later chapters this brief outline is filled in with details of actual soils and the plants most usually found on them.

The last of the three major groups of factors influencing vegetation, the *biotic factors*, should logically include the effect of the plants them-selves on each other, since they are living organisms, but it is usual to limit the phrase to mean the effect of animals only, ranging from the worms in the soil to man himself with his axe and plough. In the British Isles, and in other thickly inhabited countries, biotic factors are, of course, of prime importance. The felling and planting of trees, the cultivation of the soil, and the grazing of cattle, sheep and rabbits, have between them drastically modified the vegetation of the whole country, with the exception of those mountain, moorland, and sea-coast areas that are unsuitable for exploitation.

Such, then, very briefly, are the three main groups of factors controlling the type of plant community in any particular area. What is a " plant community " and how can the different types be distinguished and described? During the last fifty years or so, ecologists have evolved a fairly generally accepted system of classification and naming of plant communities, which it will be useful to bear in mind when reading later chapters in the book.

In the first place, a *plant community* is the general term for " any collection of plants growing together which has as a whole a certain individuality " (Tansley, 1946). Thus an isolated patch of bog asphodel on a Surrey heath is a community, and at the other end of the scale, so also are the whole heath lands of Western Europe, of which this particular heath is a sample. This very wide term, " community," is a valuable one, but clearly some further classification is necessary to distinguish between different grades of community.

For the largest unit, ecologists use the term *plant formation* to designate those communities which, as Sir Arthur Tansley says, can be recognised as " fundamentally distinct *types* of vegetation," for example, tropical rain forest and desert communities. Within these formations are smaller communities dominated by one or more particular species. When one species is dominant, as the oak in an oakwood, the community is called a *consociation*; when two or more species are equally important in the community, as in a mixed oak-ash wood, it is called an *association*. Individual consociations are referred to by names formed by adding " *-etum* " to the stem of the generic name of the dominant species, followed by the specific epithet in the genitive case; thus woodland dominated by *Quercus robur* is called *Quercetum roboris*, and so on. This terminology may appear unnecessarily pedantic, but, like the Latin names of individual plants, it has the great advantage of international currency and is now well established.

The " smallest " category of plant community is the *plant society*. This term is used to describe local communities, within an association

Plate *1*. Fruiting head of Goatsbeard, *Tragopogon pratensis*. Surrey, August

a
C. W. Newberry

b
A. Jackson

Plate 2a. Bugle, *Ajuga reptans*. Derbyshire, May
b. Betony, *Stachys officinalis*. Yorkshire, August

or consociation, formed by one or more species other than the main dominants. On a sand-dune, for example, dominated by marram grass, there may be local societies of sea-holly, or ash may form such societies in what is predominantly an oakwood.

One last point on the structure of plant communities must be mentioned here. In all except the very simplest, two or more layers or *strata* can be distinguished, each with its own characteristic life-form and species. In a wood, for example, there are four main strata, consisting of mosses, herbs, shrubs and trees. This phenomenon of stratification is dealt with in more detail later.

So far we have considered plant communities only as static " objects," existing at one point in time. It is, however, of the very essence of such communities that they are liable to change, slowly or comparatively rapidly, giving rise to what is known as *vegetational succession*. This phenomenon has been intensively studied, both in this country and elsewhere, during the last fifty years, and a rather elaborate terminology has been evolved to describe the different types and stages of succession. Tansley in his *Introduction to Plant Ecology* (1946) gives an admirable account of these terms and I cannot do better here than summarise what he says.

In the first place, two distinct types of change occur, which can be most easily understood by two simple examples. Imagine a section of cliff collapsing and forming a mass of bare soil at the foot. Plants will soon begin to grow on this soil and a sparse covering of vegetation will gradually develop. In course of time some of the first invaders, which were particularly fitted to act as pioneers, will be replaced by other species which cannot flourish until a certain amount of humus has been provided by the decaying remains of their predecessors. Still later, further species will establish themselves as seed reaches the soil, and thus a whole succession takes place from bare soil to a final cover-ing, perhaps, of scrubby bushes or small trees. Such successions, which are independent of changes in the environment, are called *autogenic* and any particular example is called a *sere*. In contrast, successions due to alterations in the environment are called *allogenic*. For example, on a chalky soil, supporting plants which are tolerant of lime, there may occur a gradual change-over to plants, like ling or heather, which are lime-haters. This is due to the leaching-out by rain of the soluble salts in the upper layers of the ground, leaving behind a neutral or even an acid soil instead of an alkaline one.

These two types of succession, autogenic and allogenic, are, of course, often in operation at the same time, and even in " pure " autogenic successions, the gradual increase in humus from the decay of plant remains introduces, in one sense, an allogenic factor.

The idea of a sere leads, inevitably, to the idea of an " end-point " to the changes involved—an end-point at which stability is reached and no further change takes place. Such an end-point is called a *climax*, and the community that has developed is called a *climax community*. Climax communities are in equilibrium with the environmental conditions as they exist at that particular time. For each particular type of climate there is a *climatic climax* community which will terminate every sere in the absence of any factor preventing the development of the climax. For example, deciduous woodland is the natural climatic climax in England, and evergreen forest in the tropics. Frequently, however, edaphic or biotic factors are present which interfere with this natural succession to the climatic climax and a quite different type of community is produced. Thus chalk grassland may be so heavily grazed that woodland cannot develop on it and the grassland is termed a *biotic climax*; or, again, soils too wet for tree-growth may remain as bog, which is an *edaphic climax*.

Other terms are also in use to describe different types of sere and climax, which are fully explained in Tansley (1946), but the very brief sketch I have given includes the basic concepts necessary for the understanding of vegetational succession.

3. Species and Evolution (M. W.)

So far in this chapter we have talked of the behaviour and characteristics of plants either as individuals or as members of a community. We must now turn to consider the groups of individuals which we call *species*. Everyone has a rough idea what he means by a different kind of plant; the gardener distinguishes between the turnip and the beetroot, the countryman between the oak and the ash, and his children between buttercups and daisies. We can very roughly say that in these groupings, made by man from earliest times, are to be found the botanist's genera, and sometimes his species. The genus, therefore, often represents that degree of appreciation of difference which is sufficient for ordinary day-to-day purposes (thus " buttercup " or " oak " correspond, more or less, with the botanical genera *Ranunculus* and *Quercus*); when the forester studies and cultivates different *kinds* of oaks, however, or the enquiring child notices different *kinds* of buttercup, we then get the concept of *species* within a *genus*.

These concepts, then, of individuals belonging to different " kinds " or species, and several species representing different " kinds " in a larger group, or genus, are as old as human civilisation; but it was not until about the sixteenth century that the naming and classification of these units was seriously undertaken. Our modern system of " binomials " (see p. 17), due to the Swedish naturalist Linnaeus, is a little over 200 years old, his *Species Plantarum* having been first published in 1753. With this background of antiquity, it is not surprising to find that the term " species " has meant different things to different ages. To primitive man it usually meant a kind of organism that one could, or could not, eat or use; to Linnaeus and his contemporaries it meant an entity specially created by God as one part of His marvellous creation; whilst to the modern botanist, who accepts the theory of evolution, it may appear to be a conveniently labelled box in which to fit, more or less successfully, groups of variable individuals at a particular point in time.

To understand a little of modern ideas on the species, therefore, we must say a word about evolution. The theory of evolution states that the diversity of organic life in the world to-day has evolved or unfolded from simple beginnings throughout many millions of years of evolutionary history. Thus, as we shall see later (Chap. 4), the flowering plants, which now dominate the world's vegetation, did not exist at the time of the vast forests of the coal-measures; they represent, in fact, a recently evolved and specialised group of organisms, whose success has gone hand in hand with a decline in more primitive groups, such as the ferns, clubmosses and horsetails. Strictly speaking, the idea of evolution is at least as old as the ancient Greeks, but it is always now associated with the name of Charles Darwin, because it was Darwin who first (in his famous *Origin of Species*, 1859) assembled a vast mass of data in support of the theory, and, what was most important, suggested a mechanism by which evolution might work—the natural selection hypothesis.

The selectionist explanation of evolution is that the offspring of organisms differ from each other, and that some of this variation is hereditary; that all the offspring cannot survive; that some, possessing particular characters, are more suited to survive than others; that, on the whole, these therefore do survive, to produce more offspring possessing (to a variable extent) the particular adaptive characters advantageous to the parents. In this way, through the operation of intrinsic heritable variation (which Darwin admittedly did not understand), the struggle for existence, and the laws of chance, each succeeding generation is kept closely "adapted " to its often subtly changing environment. By its nature, natural selection is a slow process, in all except the " lowliest " organisms (such as the bacteria) which reproduce at an enormously rapid rate; and we cannot therefore expect to see it demonstrated before our eyes. Nevertheless it is probably true to say that the majority of biologists to-day accept the selectionist view of evolution, at least in so far as it concerns what Darwin called " the origin of species."

With this view goes knowledge, not available to Darwin, of the nature of variation, knowledge which has accumulated during the present century at an amazingly rapid rate, blossoming forth as the new twin sciences of *cytology* and *genetics*. At this point we must look at this new knowledge, for some appreciation of it must underlie all modern ideas on species and evolution; but here is a very real difficulty, for a full explanation of the subject requires a book, and a brief sketch may be more tantalising than useful. Nevertheless the attempt must be made.

Cytology is literally concerned with the structure and function of the *cells* of the plant or animal; but it has come to have the specialised meaning of the study of the discrete microscopic bodies called *chromosomes* which exist, definite in shape and number, in the nucleus of every living cell. These chromosomes are the bearers of hereditary factors or genes, handed down from parent to offspring; and what we know of the behaviour of chromosomes, their mode of division, and their sub-microscopic structure, enables us to explain (in part, as in all scientific explanations) the results we observe in experiments and observations in the related field of *genetics*. Indeed the marriage of observational cytology with experimental genetics is one of the more impressive triumphs of pure scientific knowledge in this century. As a result of this work we now know that in an ordinary (" diploid ") higher plant or animal (including ourselves) every body cell has two more or less similar sets of chromosomes, one originally contributed by the male parent and one by the female. At some stage previous to the formation of sex cells in all higher plants, this double (diploid) set of chromosomes is reduced to a single (haploid) one, so that the male and female sex cells which eventually fuse each contribute a single set of chromosomes to give a new diploid individual. Normally, in the " reduction division " or " meiosis " preceding the formation of sex cells, there is an interchange of portions of similar chromosomes, so that genes derived from the male and female parent are re-combined in new ways in the resulting sex cells, and therefore in the next generation. Looked at from the genetic point of view, this reshuffling of the material basis of heredity is one of the most important attributes of sexual reproduction, the other being, of course, the more obvious one of bringing together hereditary material from two different individuals. The significance of these processes is that the constantly arising genetic changes in the chromosomes (" mutations ") have thereby a chance of " moving " through the generations of chromosomes in different combinations; and it is the different combinations of the genetic material which cause the inherited variation which we see between, for example, different children of the same family, or the seedlings from a single plant.

When we consider the variation, however, it is clear that much of it has no genetic basis; for example, portions of the same plant (possessing therefore an identical genetic make-up) grown in favourable and unfavourable conditions will, as every gardener knows, look very different in many respects, most strikingly perhaps in size.

These effects, which we can attribute to different environments, are not directly inherited at all; though for various reasons, and in certain cases, they may appear to be on a superficial view. This type of variation, due solely to the direct influence of the surroundings on the plant or animal, must be distinguished from genetically-based variation revealing itself (in the most obvious cases) whatever the environment; good examples of the latter are the occasional white-flowered variants of plants such as bluebells and violets. It is only variation of this *genetic* type which can be inherited.

Let us consider now the example of a population of some plant in a pasture subjected to heavy grazing by sheep. Unless our species is very unpalatable, it will be grazed with the abundant pasture grasses; this may mean, at the best, that it is prevented from flowering by losing all its erect flowering shoots, or at worst (if it is a loose-growing plant) that it is killed by losing so many leaves that it cannot make enough food. Under such conditions, clearly, those species will be favoured, and will spread, which are (1) unpalatable to the grazing animals (e.g. buttercups in a cattle pasture), (2) dwarf-growing, often " rosette " perennials, spreading vegetatively rather than by seed (e.g. the daisy), (3) tiny, or more less prostrate plants, whose inflorescences escape damage and thus manage to set seed. In this way we may picture the composition of a piece of grassland changing after grazing begins, the well-adapted species increasing in abundance at the expense of the others. Now clearly this does not only apply to different species, but also, though perhaps more subtly, to different individuals of the same species; under such conditions, for example, a rather loose-growing daisy will, on the whole, be less successful than one with a neat, compact rosette. In this way, those individuals best fitted to the particular conditions gradually come to represent the species in the particular place; and in different conditions the same species may be represented by individuals with genetically different habits of growth. Such specially adapted types within species characteristic of particular kinds of habitats are called *ecotypes*; and many common species show this kind of adaptive variation, although the more abundant and ubiquitous the species, the less likely it is that the boundaries between different ecotypes will be distinct. In cases where a species shows this more continuous type of adaptive variation we speak of an *ecocline* with respect to the particular character (e.g. dwarfness) shown.

This kind of adaptation will clearly bring about, in two populations of any species isolated from one another and under different conditions, a gradual divergence, which may eventually be so obvious that we call the two populations different species, or at least different subspecies. In this way we imagine that many species-pairs have arisen from an original " parent " species. The *isolation* of populations seems to be the important element, and isolation may be due to ecological differences, to mere distance (geographical isolation), or to barriers such as a sea or a mountain range. It seems possible that populations prevented from interbreeding will gradually come automatically to differ in genetic make-up—and therefore, usually, in appearance— quite apart from any " selection pressure " from different environments. Here, then, is one method by which we believe new species arise. It is clearly a gradual method of speciation, and its end-point may often be two or more species so different genetically from each other that they are not capable of producing fertile offspring when brought together and crossed, or allowed to hybridise naturally. It should, however, be noted that there are a number of well-known cases of two obviously different plants, which have always been treated as different species, whose hybrids are fully fertile (e.g. the red and white campions, see p. 190); such cases are very interesting to the student of speciation.

The opposite situation—that is, two groups of plants so similar in appearance that they can only with difficulty be distinguished, yet unable to give fertile offspring with each other at all—tends to result from the other main method of speciation, which we call *polyploidy*. Many groups of closely-related species have chromosome numbers which bear a simple multiple relation to each other—the docks (*Rumex* spp.), for example, have various numbers in multiples of 20, from 20 to 200—and the higher (polyploid) members of such a series are thought to have originated from the lower by a process of chromosome doubling. In simple cases, known as *auto-polyploidy*, a tetraploid arises directly by doubling of the diploid chromosome set; in the more complicated, but in nature much more important, cases of *allo-polyploidy*, a more or less sterile species-hybrid is converted, by chromosome doubling, into a fertile new " hybrid species." Both auto- and allo-polyploids can be produced artificially, by use of the chemical colchicine, which has the effect of delaying cell division whilst permitting nuclear (and therefore chromosome) division. Artificial allo-polyploids, produced by hybridisation and doubling, are in every way " good " new species; and indeed we now know in some cases, or have good reason to suspect in others, that some of our commonest and most important plants have originated in nature in some such way as this. The abundant and successful weed, annual meadow grass (*Poa annua*), and the important, variable cock's-foot grass (*Dactylis glomerata*) are two good examples of species almost certainly of allo-polyploid origin; and in the famous case of the cord-grass, *Spartina townsendii* (see p. 174), we can give time and place to the origin of the new species in nature!

These then seem to be the two main types of speciation at work in nature, and both gradual speciation and polyploidy are obviously important in our flora. We have, however, a good many examples of species or species-groups which do not fit these categories. In these cases there is usually known to be partial or complete *apomixis* (reproduction not involving sexual fusion), and the " species " here is a very different type of group from the normal sexual species (see p. 122).

We might finally consider briefly what is the relation of these methods of speciation to evolution as a whole. Clearly, knowledge of how species may arise does not explain the whole course of evolution. We have, as it were, only learnt a little about the branching of the finest twigs of the tree, and nothing of the origin of the main branches, not to mention the trunk (or trunks, for we do not know whether life had one or many origins). It is, however, reasonable to assume that gradual speciation under selection in different and changing environments would be applicable throughout evolution, and that some at least of the species of to-day are the starting-point of the genera and families of the future. The role of polyploidy in large-scale evolution would seem to be much more restricted, if only because it has the appearance of an evolutionary cul-de-sac ; for chromosome sets may be doubled, but not halved again by any comparable process. We should, perhaps, in all humility, confess that the main evolutionary process in the plant kingdom is still shrouded in considerable mystery, and indeed seems likely to remain so for many years to come. It is a fascinating mystery, and one on which speculation is easy, but concrete evidence hard, or, by the nature of the problem, in some cases impossible, to obtain; its consideration at present lies as much in the realm of the philosopher as of the scientist.

Plate 3. Fruiting spikes of Lords-and-Ladies, *Arum maculatum*. Hertfordshire, August

Plate 4. Bluebells, *Endymion nonscriptus.* Kew, May

CHAPTER 4

HOW OUR FLORA CAME TO BRITAIN
(M. W.)

THE MODERN GENERATION is very familiar with the concept of change; indeed, one might say that to-day the impermanence of our surroundings, including the society we live in, is all too obvious for many of us. In the same way, the awareness of the importance of change in the understanding of our vegetation, whether over long geological periods or shorter historical ones, is a relatively recent development in botanical thought. Of course, fossil animals and plants have been known for a long time, and men have speculated on the evolution of living organisms since the days of Greek science; but until Darwin had firmly established the theory of organic evolution in the middle of the last century, there was little advance in our understanding of the past history of plants and animals in the world.

The last century has seen, firstly, a great increase in our knowledge of fossil plants, and then, in more recent years, our knowledge of vegetation changes in time which can be measured in thousands rather than in millions of years has made a spectacular advance. Of course, there are still enormous gaps in this knowledge; but it is at least possible now to present some kind of picture, however inadequate, of the history of the plants which have grown throughout different periods in the area where the British Isles are now situated.

Everyone in England is probably familiar with fossil plant remains in coal; and indeed we all know that coal is entirely composed of the remains of vast forests of trees which flourished in swamps 300-400 million years ago, in the so-called Carboniferous age. The study of these primeval forests, as revealed in their preserved fossils, is a very absorbing one; but we must limit our remarks to more recent geological periods, for the flowering plants were quite unknown in the

Carboniferous. But the origin of our modern flowering plants is still a tantalising mystery; all we know is that there were none (or rather we have no fossilised remains) before the Jurassic period—the age of Oolitic limestones—whilst by the Cretaceous (our chalk), rocks in several different parts of the world contain many different flowering plant fossils. Some of these earliest fossil flowering plants are, for example, relatives of the plane trees (*Platanus*) and the magnolias which we grow in our gardens to-day; but the record is very imperfect, and there must have been many other kinds.

By the beginning of the Tertiary epoch (Eocene), however, we find that, preserved in the London Clay deposit, we have abundant evidence of a rich flora, very similar to that of a tropical forest of to-day (see Reid & Chandler, 1933). It is rather difficult to imagine a tropical swamp-forest where now sprawls the enormous mass of London. Throughout the long ages represented by the Tertiary rocks our flora seems, from the very inadequate records, to have changed gradually from a " tropical " one to one containing familiar north temperate plants, until by the end of the Tertiary, we have, in the Cromer Forest Beds, many records of plants which we are familiar with in Britain to-day, together with a few, such as the silver fir (*Abies*) and the water-chestnut (*Trapa natans*), not now native in these islands.

Then occurred the event—or rather the long-drawn-out series of events—which must have wiped out the vast majority of all plants and animals from the face of Britain. This was the onset of the ice age, when the polar ice-cap gradually extended its area southwards until, at its height, enormous and more or less continuous sheets of ice occupied most of present-day Britain except the south of England. It is not strictly correct to speak of *the* ice age, for we have evidence of several advances and retreats of the ice—at least four separate ice ages, in fact—and, between each, in the so-called inter-glacial periods, there flourished again some kind of temperate vegetation where the ice had once held sway, and was to return. As yet we know very little of the inter-glacial periods; what we do know is that the ice finally began to retreat some 15-20,000 years ago, and, as it retreated and the climate became less arctic, the vegetation covering land south of the ice moved up in waves or zones into the area left bare by the retreating ice.

The detailed study of vegetational changes since the ice age—the so-called Quaternary period—is one of the most fascinating and most recently developed branches of botany, which links up the work of the geologist, the archaeologist and the historian in the joint unravelling of a complex story. (Even the nuclear physicist has recently been brought in, as we shall mention later.) The botanist studying the Quaternary period does so by investigating what he calls " sub-fossil " plant remains, found in peat and lake mud deposits. Peat (see Chap. 9, p. 132) is made up of only partially decomposed plant remains— in fact, coal can be looked upon as the end-product of peat subjected to enormous pressures in the earth over millions of years—and in peat are preserved, not only parts of plants easily visible and with care and practice identifiable (such as whole " bog-oaks," or twigs, leaves, seeds, etc.), but also a wealth of microscopic pollen grains. The rapid success of this study has been due to the convenient fact that the outer shells of pollen grains are not only extremely resistant to change and decay, but also that many of them possess characteristic shapes and precise patterns of sculpturing, by means of which it is possible to find out by microscopic study what kind of plant they came from. Now most people know, and hay-fever sufferers are pain- fully aware, that different kinds of plants produce very different amounts of pollen. Some, like the grasses and most forest trees, produce masses of dust-like pollen blown by the wind; whilst most plants with showy flowers have only a comparatively small amount of pollen which is transferred to the stigma of another flower usually through the agency of an insect visitor. Clearly it will be the wind- pollinated plants which contribute the vast majority of the pollen grains to the peat record, for their dust-like pollen is everywhere in the air. It is found, by taking samples of peat from different levels in a raised bog and treating them in such a way as to separate out the many pollen grains, that the great majority of these grains do not belong to the plants which made up the main mass of the peat (e.g. bog-moss, cotton-grass) but to a few common forest trees, which must therefore have been present abundantly in the vicinity of the growing peat-bog. From the numbers of each kind of tree pollen, it is possible to prepare a " pollen diagram " (Fig. 1, p. 50) showing change in relative proportions of different tree pollens from the lowest, and therefore oldest, layers of the bog to the most recent upper layers. The detailed interpretation of such a diagram is not easy; but at least its general implications are fairly clear. It is reasonable to assume that if there is a relatively large amount of the pollen of, say, pine at a

50

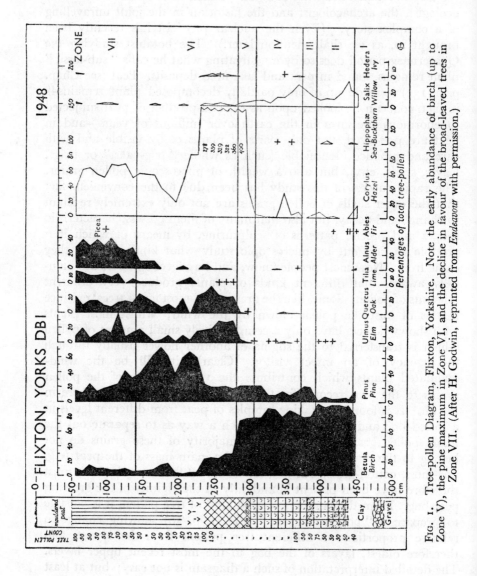

FIG. 1. Tree-pollen Diagram, Flixton, Yorkshire. Note the early abundance of birch (up to Zone V), the pine maximum in Zone VI, and the decline in favour of the broad-leaved trees in Zone VII. (After H. Godwin, reprinted from *Endeavour* with permission.)

G. Atkinson

Plate 5. Yellow Dead-nettle, *Galeobdolon luteum.* Kew, June

Plate 6. Lesser Celandine, *Ranunculus ficaria.* Cheshire, March

certain stage in the history of the bog, and at a later stage the pine declines, until in the upper layers there is hardly any at all, then the amount of pine in the forest surrounding the bog must have changed very markedly over the period of time represented by the growth of the bog. The relative amounts of the tree pollens give us a rough guide to the composition of the forests surrounding the bog as it grew slowly over thousands of years.

The next step, of course, is to try to decide the approximate ages of the different levels represented in such a pollen diagram. This is 1.o easy matter, and to explain it fully would require a book in itself; but the general principle has been to correlate the pollen diagrams with evidence of all kinds from other studies (such as that of archaeological remains in the peat), until the jigsaw puzzle begins to fit together. Once it begins to fit, of course, we know we are on the right lines, and gradually an impressive composite picture is built up, based on very many pollen diagrams, and all sorts of other evidence, over as wide an area as possible. In recent years great strides have been made with the " C14 " method of determining the age of plant and animal material by estimating the proportion of radioactive to ordinary carbon in it. Dates estimated by this method are now generally in good agreement with those estimated in other ways, and the method has progressed beyond the experimental stage.

Most of the pioneer work in pollen analysis was done by the Swede, von Post, who had worked out, by the year 1916, the main sequence of changes in the vegetation cover since the ice retreated from northern Europe, and had provided with it an approximate time-table. It is, however, only very recently that we have begun to get a picture in more detail of the earlier stages at the beginning of the retreat of the ice; and in order to present the tale in chronological sequence, we should begin with this most recent work. We have seen that the great majority of pollen grains found in most layers of bog peat are those of a few familiar forest trees. If, however, the lowest layers, which may be muds rather than peat, of certain deposits are investigated, it is found that tree pollen is absent from them, and instead there is a characteristic assemblage of pollen of herbaceous plants. From such deposits valuable evidence is being gradually accumulated as to the kinds of plants which were growing in various parts of north-west Europe, including what is now Britain, between the period when ice covered most of this region and the latter stage when the trees had recolonised the ground and were forming more or less complete

forests over it. Professor Sir Harry Godwin, who has contributed most
to the understanding of this whole subject in Britain, has given (1949)
an interesting account of this early vegetation of the Late Glacial
period, and has drawn attention to some of the more surprising
discoveries which are being made in this field. When we think of
Britain in this period, we must remember that our island was joined to
the continent of Europe (where the English Channel is now), and
Ireland to Britain, not only during the last ice age, but also for a
considerable time after the final retreat of the ice from our region.
It is therefore reasonable to suppose that much the same kind of
vegetation was present over that part of northern Europe, including
what is now the south of England, which was not covered by the ice at
its greatest extent. At the edge of such an ice-sheet, we can imagine a
sort of treeless tundra similar to that found in the arctic to-day, where
the gravelly or rocky ground has a sparse covering of low-growing herbs
and dwarf shrubs, with mosses and lichens. The climate, we can
imagine, was very cold, and may also have been very dry—there is at
the present time an arctic desert, with salt-lakes, near to the permanent
ice-sheet of Greenland—and under such conditions the growth of trees
would be quite impossible. Then the slow retreat of the ice began;
it did not go smoothly, but made at least one re-advance before finally
retreating. Roughly between 20,000 B.C. and 8,000 B.C. we can say that
over much of north-west Europe, including Britain, there was a mixed
and rather open type of vegetation into which scattered trees of birch
and pine were advancing as the climate gradually improved. In this
Late-Glacial countryside enormous areas of country were covered by
low-growing plants; the grasses and sedges must have been abundant,
for their pollen is present in enormous quantities. In addition to these
abundant herbs, a remarkable variety of other plants were present,
some of which occupy very restricted habitats in present-day Britain.

A most important group are the mountain and moorland plants,
such as the mountain avens (*Dryas octopetala*), the dwarf willow (*Salix
herbacea*) and the crowberry (*Empetrum*), which to-day occur abundantly
in arctic tundras as well as on our mountains (see p. 114). Another
important group are the marsh and water plants, such as the pond-
weeds (*Potamogeton* spp.), the reedswamp sedges (e.g. *Carex rostrata*), and
the meadowsweet (*Filipendula ulmaria*, Plate 24, p. 177). We can well
understand why this kind of plant was so abundant. The enormous
areas of ice would produce, every summer, floods of " melt water,"

and we can envisage impassable marshes, with luxuriant vegetation, on the lower ground. The existence of such rich feeding-grounds no doubt explains the occurrence at this period of very large herbivorous mammals—the bison in Denmark, the elk in Britain, and the giant Irish deer—which became extinct in our regions in the later forest period. A third interesting group of Late Glacial plants are those which to-day occur chiefly or almost wholly as weeds, dependent on human cultivation or other activities. It had previously been suspected by many botanists that some at least of the common weeds had natural habitats in this country such as sand-dunes, sea-cliffs and river-banks; but the extent to which weeds are proving to be old and respectable members of the flora is really rather surprising. Even such an apparently pure weed as the cornflower (*Centaurea cyanus*) has been found (as pollen) in Late Glacial deposits in Breckland! But we shall have more to say about this in Chapter 13.

In addition to these three categories, there is another group of plants, less easy to define, which were represented in these Late Glacial floras, but are now rare or local plants of chalk and limestone. These are discussed in Chapter 7.

The picture, then, is of an open, at least partially treeless vegetation (which has been called " park tundra "), rich in many different kinds of plants, over which large grazing animals roamed, and also primitive man, their hunter. This type of vegetation lasted several thousand years, but was gradually replaced by woodland as the climate improved. The first trees to make a more or less closed woodland were the light-seeded birches (*Betula*); at the present day, the birch forest goes farther north in arctic Europe than any other tree. These gradually gave way to pine (*Pinus sylvestris*), until, in the so-called Boreal period, most of lowland England must have been covered with pine forest, in which the hazel (*Corylus avellana*) was abundant. This Boreal pine forest period we can date at approximately 8,000-6,000 B.C., and its end came with a rather sudden change of climate in Britain, from cold and dry to mild and wet. This is the beginning of the Atlantic period, when the familiar broad-leaved deciduous trees, oak (*Quercus*), elm (*Ulmus*) and lime (*Tilia*) appeared in quantity and began to replace the pine. The origin of the English Channel is usually thought to date from about this time—indeed, the suggestion has been made that the climatic change resulted partly from Britain's becoming an island. In the mild wet Atlantic period the oak forest flourished in England, and in the waterlogged valleys alder (*Alnus glutinosa*) was very abundant. The Boreal pine forests were now more or less restricted to the north of Britain.

So far, there is little or no evidence of any serious effect by man on the vegetation. Towards the end of the Atlantic period, however, the Neolithic human culture spread, to expand very considerably in the drier Sub-Boreal period (c. 3,000-2,000 B.C.) which followed. On the Wessex chalk and in the Breckland area in particular, New Stone Age men may well have cleared considerable areas of forest from the thinner soils, on which their task was comparatively easy (see Chaps. 6 & 7). The drier climate would no doubt have helped this process. From that time onwards, though naturally with many fluctuations in intensity, the British countryside has been affected, indeed shaped and created, by man, until to-day it is difficult to find a natural plant community undisturbed either directly or indirectly by human activity.

During the Sub-Boreal period the pine forest succeeded in coming back to some extent, spreading, for example, on to the edge of the now drier fenland and bog; but a further onset of wet conditions in the Sub-Atlantic (c. 500 B.C.) reversed this process. As we shall see in Chap. 5, the only certain direct descendants of the vast Boreal pine forests of 7,000 years ago are to be found in the remnants of native pine forest in the Scottish Highlands. Compared with these ancient remnants, the southern English beech forests are very recent arrivals, for the pollen of beech is not present in quantity in the peat record before the Sub-Atlantic period.

This fascinating story of change gives us a new light on the history of our plants. For one thing, we become aware that, quite apart from change due to man's activities, vegetational change, both small- and large-scale, has always been going on, and must still be taking place. Even without the direct aid of man, new plants will still be arriving in this country, and native plants spreading or declining in response to subtle changes in climate and soil. Indeed, viewed in the light of the history of the last 20,000 years, the present day may well be situated in the fourth inter-glacial period between two ice ages—the ice has, geologically speaking, only just left, and may well return again "fairly soon"! So practically the whole of our flora is a recent immigrant one, poor in species compared with the vast wealth of the ancient tropical floras in other countries, which were not affected by the ice age. When we first begin to learn to distinguish British plants, we are naturally impressed by the size of the task; but we can at any rate console ourselves that almost anywhere else in Europe, Asia or North America the task would be very much harder! It seems clear that we lost a good many plants during the ice ages, which, like the silver fir (*Abies alba*), have not succeeded in returning to the flora— except as garden plants or planted trees, which may grow quite well in our present-day climate. Their absence we may look upon as due to a historical accident—chiefly no doubt the accident of Britain's becoming an island. This is even more true of Ireland, whose flora and fauna show striking absences. Everyone is familiar with the fact that snakes are absent there, and there are several equally impressive, though less well known, gaps in the Irish flora.

It is possible to divide the British flora into a number of groups according to the distribution of the species concerned within and outside the British Isles. The main divisions (cf. Matthews, 1937) are as follows:

1. Mountain plants, occurring in the Arctic, the Alps, or (in the majority of cases) in both these regions (arctic-alpine). (See Chap. 8.)
2. Northern moorland or forest plants, common in northern Europe. (See Chaps. 5 and 6.)
3. " Continental " species, common in central and/or eastern Europe, rarer in the west. (See Chap. 6.)
4. " Atlantic " (incl. " Lusitanian ") species, more or less confined to the regions of the Atlantic seaboard of Europe.
5. Wide-ranging species occurring throughout Europe and much of Asia, or even more widely distributed.

Groups 3 and 4 in particular are much subdivided in Matthews' treatment. It is significant that we have hardly any *endemic* plants (i.e. confined to the British Isles).

We can appreciate now something of the reason for these groupings in the flora. The mountains shelter the relics of a " periglacial " flora once widespread at the edge of the ice-sheets. The northern moorland and forest remnants likewise still have relics of the Late Glacial and Boreal floras, though a good many of these plants are also continental plants of limestone and chalk to-day. The history of the Atlantic element in the flora is less clear; but we can at least say that plants of this type must have had their greatest opportunity of spread in the Atlantic and Sub-Atlantic periods. The rarity of endemic plants emphasises how recently our flora has re-immigrated; there has been no time for speciation or subspeciation within the British Isles, except in a few apomictic groups, such as the hawkweeds (*Hieracium*) and the whitebeams (*Sorbus*), and in one or two others.

We have seen how our flora was subjected to catastrophic changes long before the human race took any hand in shaping it. It is clear that man is now the over-riding influence, and future changes in the composition of the flora are largely dependent upon his actions. As we shall see in Chapter 13, some recent human introductions have been remarkably successful; so that even with the spread of industrial conditions and the destruction or taming of so much of the countryside, not all is loss for the student of the British flora. Indeed, as we have endeavoured to show in this book, the processes of change are them-selves well worthy of study; and the activity of man is one factor among many which we must take into account. For an introduction to further reading on the subject of the history of the British flora and vegetation, Pennington (1969) is much the most suitable book. In it can be found references to all the other books and papers, such as the important standard work by Sir Harry Godwin (1956).

CHAPTER 5

WOODLANDS AND HEDGEROWS
(J. G.)

Most people would describe Britain today as a well-wooded country. Our Ordnance maps are pretty heavily mottled with green. But if maps had existed five thousand years ago, sheet after sheet would have shown nothing but woodland; only where there were high mountains, fens, bogs, and perhaps chalk downs, would white patches have indicated areas bare of trees. These ancient forests have shrunk to their present patchwork pattern for two interconnected reasons. If civilised man was to establish himself in Britain he must have clear land on which to grow his crops and to lay out his villages; and he must have timber to build his houses, vehicles and ships, and to make his tools and furniture. But the story is not quite as simple as this. The first need—the need for clear land—could be met by destruction without replacement, but increasing need of timber could not be satisfied indefinitely by cutting down existing trees and waiting for new ones to grow; so man began to reverse the process he had begun and to plant where he had felled. Our woodlands have been profoundly affected by this dual policy, more especially in their reduced extent, but also, to some degree, in the kinds of trees, shrubs and herbs that make up those that are left. Today, there are very few woods that have escaped direct interference by man, by felling, planting and coppicing, and even those that have done so have been more or less modified by the cultivation of surrounding land; for example, by the introduction of weeds into their natural ground flora.

Nevertheless, despite this interference, " tho' much is taken, much abides," and many present-day woodlands, even those entirely composed of planted trees where native species have been used on their appropriate soils, give a very fair idea of what our virgin forests must have been like in prehistoric times.

57

All plant communities arrange themselves in a series of layers, like the stories of a house. Even in grassland close-cropped by sheep there are usually two layers; the short grasses and the rosette plants form one, and the larger herbs not eaten by the sheep a second. Woodland has the most complex structure of all communities. There are at least four layers—containing mosses, herbs, shrubs, and trees respectively —and each layer has its own little " climate," differing from the others in light intensity, and in the concentration of carbon-dioxide, oxygen and water vapour in the air.

In this book we are concerned primarily with the layer composed of herbaceous plants—often called the " field " layer. I will not attempt, therefore, the fascinating task of describing each type of woodland in turn—oakwoods, beechwoods, birchwoods and the rest —and of showing the relationship between the different layers in each type; separate books in this series are to be devoted to British woodlands, to trees and shrubs, and to mosses. All I can do is to focus attention on a few of the most characteristic species in the field layers of the different woodland types and to try and show them against a background of their environment as it changes with the changing seasons.

OAKWOODS

Spring and summer wear very different aspects wherever plants are growing. By roadsides, the white ribbon of hedge parsley gives way to the darker flush of the woundwort, and along the stream nearby purple loosestrife ousts the golden drift of kingcups. But it is in a woodland, where the coming of the green canopy turns spring morning into summer twilight, that this universal change is most complete. Springtime, when sunlight can find its way to the floor of the wood through the bare branches overhead, is the period of maximum growth—the period when the majority of woodland plants shoot and flower, and when the few summer flowers develop their leaves and make their food for the shadowed months to come.

There are two species of oak in Britain, the pedunculate oak (*Quercus robur*) and the sessile or durmast oak (*Q. petraea*, formerly known as *Q. sessiliflora*). *Q. robur* has stalked acorns and almost sessile leaves; *Q. petraea* has sessile acorns and stalked leaves. Though the two species overlap and hybrids between them are recorded, the pedunculate oak is mainly a species of moist, heavy loams and clays, while the durmast oak favours lighter, more sandy, shallower and drier soils. Pedunculate oakwood, with hazel dominant in the shrub layer, so familiar in the heart of England on the deep midland loams, is the most characteristic British woodland. It is the " climatic climax " (see p. 40) that would develop over most of our heavier land if man were to cease cultivating the soil.

The field layer in oak-hazel woods today is profoundly influenced by the coppicing of the hazel, which is carried out every ten to fifteen years. Just before coppicing, the dense shade of the hazel prevents all but a few flowers developing beneath it. Two or three years later dormant rootstocks and seeds have sprouted again and a fresh supply of seed has invaded the now lightly shaded ground. This is the moment to see at its best the mosaic of flowers that paves an English oakwood, the windflowers, the bluebells and the primroses whose

> " forme and . . . infinitie,
> Make a terrestial galaxie,
> As the small starres doe in the skie."
> (John Donne)

One of the earliest flowers to respond to the spring warmth is the lesser celandine (*Ranunculus ficaria*, Pl. 6, p. 51), whose gleaming gold challenges the pale sunshine that gave it birth. Its life above ground is short. Soon after Christmas green buds push through the ground, and by May or June the leaves, flowers and fruits have withered away. Below the soil, however, there are thick roots now stored with food against the spring emergence of the coming year. Dog's mercury (*Mercurialis perennis*, Pl. IV, p. 35) is as early as celandine in pushing through the cold January soil and even earlier in flower, sometimes before the first of February. Its leafy stems, however, unlike those of the celandine, persist right through the summer, often forming a close carpet over many square yards. It prefers oakwoods on basic soils, and is indeed more at home in ashwoods and in beechwoods, with an open canopy and sufficient depth of soil. Even earlier to flower than either celandine or mercury is the little winter aconite (*Eranthis hiemalis*, Pl. 7b, p. 66), a garden escape from southern Europe, now naturalised in woods and plantations in many parts of the country.

Dog's mercury is not a spectacular plant, except in sheer weight of numbers and power to keep all rivals from its chosen ground. Close on its heels, however, follow the lovely white stars of the wood anemone,

> " *Dyed in winter's snow and rime*
> *Constant to their early time.*"

(John Clare)

Anemone nemorosa (Pl. 8a, p. 67) shoots at the end of February and flowers a week or two later. Like celandine, its growth period is short and by June its leaves are gone. Underground, however, there remain creeping rhizomes which not only store up food made by the leaves during their brief encounter with the sun, but enable the plant to enlarge its territory and to occupy a whole copse perhaps, or a lightly shaded woodland glade. This vegetative reproduction leads to a remarkable uniformity over large areas of the wood anemone's distribution, but colour varieties—blue and purple—do occur here and there.

Is there anything new to be said about the primrose? We have emphasised throughout the book that the life-history and ecology of many of the commonest British plants are still very little known. Professor R. Good (1944) has pointed this moral by an intensive survey of the distribution of primroses in Dorset which has revealed several new facts about the habits of this best known and loved of woodland plants. Good travelled " every road and major track in Dorset, many of them more than once," and recorded all the primrose communities that came within his view. This survey showed, as might be expected, that Dorset primroses were almost confined to woods and hedge-banks; while nowhere did they grow in hedgebanks but not in adjacent woods. Further, he found that there were two stretches of country where primroses were almost entirely absent, one running diagonally north-east across the centre of the county, and a second smaller area in the extreme east. These gaps create two main primrose areas, a large one in the west and north-west of the county, where the plant grows equally in woods and hedges, and a narrower area towards the east, in much of which it is in woods only. The key to this lies in the distribution of soils and rainfall in the county. The two " primrose gaps " coincide roughly with the presence of dry, chalky or sandy soils, while the two areas where the plant occurs are on damper clays or loams. This correlation shows that type of soil is the main factor governing the distribution of the primrose; but rainfall comes in as a modifying influence. In Dorset the highest rainfall is in the west, gradually diminishing eastwards, though high ground breaks the regularity of the gradient. Now, it is in the west that primroses are most abundant and inhabit both woods and hedges, and in the eastern area that they are more thinly scattered and often grow only in the woods. It would seem then that, in Dorset, the primary requirement of the plant is a moist habitat. This it gets in woods on clays and loams in both the east and west of the county; on hedgebanks, however, which dry out more quickly than woods, it is much commoner in the west, where the higher rainfall keeps the hedgebanks moist. Nowhere are hedgebanks wetter than the nearby woods, which explains why Good found no hedgerow primroses in areas where they were absent from woodland.

I have quoted this investigation rather fully in the hope that others will be inspired to follow the primrose path in their own counties and compare their results with Good's, and partly because it is a first-class example of the fairly straightforward (though time-consuming) type of research that is badly wanted in many other common species.

Those who live in that part of East Anglia where Suffolk, Essex and Cambridge meet are so familiar with the oxlip (*Primula elatior*) that they find it difficult to believe that it grows nowhere else in the British Isles. Within this small area it is abundant in woods on calcereous boulder clay overlying chalk, and occasionally in damp meadows.

Over sixty years ago Miller Christy (1897) made a thorough study of the distribution of the oxlip and primrose in East Anglia and put forward a theory which immediately caught the imagination of naturalists. He found that around the edge of the oxlip area there was a zone of oxlip-primrose hybrids, and he suggested that this hybrid-zone was gradually extending inwards; that the oxlip was, in fact, being slowly " hybridised out of existence " by the primrose. That this can happen in other species is shown by the history of *Medicago falcata* in East Anglia (Gilmour, 1933) and probably the red campion in Cambridgeshire (see p. 190), but Professor Valentine's work (1947b) on primroses and oxlips has shown that this encroachment of hybrids is unlikely to have taken place, and that the movement of the two species is probably governed by ecological factors such as the dampness of the soil. For an account of some interesting recent work on this problem and an assessment of the rival interpretations, see Woodell (1969).

The oxlip has the many-flowered umbel of the cowslip, but each flower is nearly the size of a primrose, though of a deeper yellow. These resemblances often betray those who do not know the true oxlip into mistaking for it the primrose-cowslip hybrid, or " false oxlip," which is quite common in many parts of Britain where the two species grow together.

Robert Bridges has vividly painted the ground flora of an oakwood:

> " *Thick on the woodland floor*
> *Gay company shall be,*
> *Primrose and Hyacinth*
> *And frail Anemone.*"

The second of Bridges' trio of woodland flowers, the bluebell (wild hyacinth in Scotland) brings in the summer, as the anemone and primrose have brought in the spring. At the end of April or in early May the three are indeed " gay company " where they grow intermixed, but in many oakwoods on light soil the carpet of bluebells leaves no room for other plants.

To visitors from central and eastern Europe British bluebell woods are a revelation; I have even heard one admit that they are a fair substitute for the flower meadows of his native Alps. The species (*Endymion nonscriptus*, Pl. 4, p. 47) is a typical "Atlantic " one, not reaching farther east than western Germany and absent even from Scandinavia. It begins to shoot just before *Anemone nemorosa*, flowers from mid-April to early June, and its leaves have turned flaccid and yellow by the end of the month. The bluebell is one of the most popular wild flowers for picking, and there have been many arguments as to how much damage is done by constant picking, and whether pulling up the whole flower stalk harms the bulb more than nipping it off above ground level. By a coincidence two experiments were started at almost the same time, one near Oxford by Mr. T. R. Peace and a second at Kew by myself and other members of the staff, to settle this controversy and to work out some sensible advice to bluebell-pickers (Peace and Gilmour, 1949). In both places plots were marked out and divided into four sections in which, respectively, (1) the leaves and flowers were destroyed, (2) the flowers were picked with whole stalks, (3) the flowers were picked above ground, and (4) the plants were left alone. After eight years of this treatment at Kew the last three sections were all just about as full of bluebells as they were before the experiments started, whereas the section in which both the leaves and flowers had been destroyed contained only a few starved plants. This behaviour might, of course, have been forecast from our knowledge of the economy of bulbous plants in general, since the prevention of seed-formation normally strengthens such plants, while the continued destruction of their leaves when photosynthesis is at its height would clearly weaken them in course of time; nor is it surprising that the removal of the base of the flower-stalk has no ill effect on the bulb. Nevertheless, here was visible demonstration of the expected, and the experiments enable one to say definitely that moderate picking (above or below ground) will do no harm, provided care is taken, especially by large parties of pickers, not to trample on the leaves.

The flowering of the bluebell ends the great spring display in our oakwoods. I have mentioned the chief plants contributing to it; but there are, of course, others. Some are equally widespread, like the woodspurge (*Euphorbia amygdaloides*), ground ivy (*Glechoma hederacea*) and the two violets, *V. riviniana* (Pl. 8*b*, p. 67) and *V. reichenbachiana* (see p. 92); others administer a stab of excitement each time we meet them. *Gagea lutea* is one of these. I well remember, when I was at school, first seeing its greenish-yellow stars, through a drenching April shower, in a wood near Malvern, and a year or two later, near Uppingham, coming upon it again, bathed in April sunshine. It is a little bulbous plant belonging to the lily family, first recorded from Somerset by L'Obel in 1570, but more frequent in the north of England than farther south. Its detailed distribution, ecology and life-history would be well worth investigation. For several years it may remain flowerless, and in every patch there is a high proportion of bulbs bearing leaves only. It apparently rarely sets seed, and reproduces itself by minute bulbils, as many as twelve or fifteen from the parent bulb.

Two other early spring woodland plants that are always exciting to find are *Helleborus viridis* and *H. foetidus* (Pl. III, p. 34), the green and the stinking hellebore. They are both grown in gardens, like their relative the Christmas rose (*H. niger*), and often, in apparently native habitats, they are relics of past cultivation; but in many English woods—especially on a calcareous soil—they are certainly indigenous. As in some other members of the *Ranunculaceae*, their petals (pure green in *viridis*, duller green in *foetidus*, with a purple border) are equivalent to the sepals of a buttercup, the petals of the buttercup being represented by little tubular nectaries hidden at the base of the drooping flowers and protected from rain by the functional petals. They are pollinated by early bees.

The moschatel (*Adoxa moschatellina*) is a widespread, but rather local, plant of oakwoods in April and May; its little head of five green flowers is so easily missed that it is often thought of as a rarity. By coining the word *Adoxa* (Greek *a* and *doxa*—lacking in glory), Linnaeus emphasised this inconspicuousness, but botanically the moschatel has belied its name, since it is now generally considered to have the distinction of being the sole representative of the family *Adoxaceae*. The specific epithet " *moschatellina* " means musk-scented, but I must confess that it takes a keener nose than mine to detect the " faint musky smell " it is said to emit when " young and moist with dew."

As the oak leaf canopy gradually closes, deepening the shade beneath, the primrose rosettes begin to take on a cabbagy luxuriance; by early June, spring has made way for summer and a crop of new flowers. There are many species in this second phase, especially in rides and clearings, but few make so bold a show as the wind-flowers and bluebells of the earlier months.

Oakwoods (and mixed woods of oak and ash) on calcareous soils share many of their summer flowers with beechwoods (see p. 68), but one or two species are particularly characteristic of this habitat. Herb Paris (*Paris quadrifolia*, Pl. IV, p. 35) and columbine (*Aquilegia vulgaris*), both flowering in early summer, and the nettle-leaved bell-flower (*Campanula trachelium*), which begins to flower in July and lasts till September, are a representative trio. Herb Paris is an intriguing plant, with four (rarely five or six) leaves arranged in a whorl beneath the single greenish-yellow flower. Like some other shade species it rarely produces fruits and must spread mainly by its underground rhizomes. Wild columbine is much more local than *Paris* and is not a native (though sometimes a garden escape) farther north than the Scottish Lowlands. It likes more open ground and I have seen it at its finest on the margins of oak-ash woods on the North-amptonshire Oolitic limestone. The bell-flower also favours these woods; it thins out towards Scotland, where its place is taken by the much handsomer species, *C. latifolia*, with large solitary blue or white flowers in the axils of the leaves.

Two interfertile species, when growing together, often produce a tangle of mixed forms known as a " hybrid swarm." A pair of summer woodland plants which behave in this way are wood avens or herb bennet (*Geum urbanum*) and water avens (*G. rivale*). Wood avens is the commoner species, but both are widespread over the whole of Britain. In woods where dry, open ground abuts on a marsh or stream-side the two may meet and a whole range of intermediates is the result, including both first-generation hybrids and back crosses with each parent. The small yellow flowers of wood avens are very different from the drooping, purplish-brown bells of *G. rivale*, and it is fascinating to try and pick out the various forms in a hybrid swarm and to judge their likely parentage with the help of Mr. E. Marsden-Jones' beautifully illustrated paper in the *Journal of Genetics* (1930).

In the summer, most of the conspicuous oakwood flowers grow in small colonies; very few spread themselves through a whole wood like the spring anemones and bluebells. A lovely exception is *Myosotis sylvatica*, a forget-me-not with a flower nearly twice the size of the common species, *M. arvensis*. There is a large-flowered woodland variety (var. *umbrosa*) of the last species which is often confused with *M. sylvatica*, but you can tell the true *sylvatica* by its more deeply divided calyx and longer style (equalling the calyx). It is rare in the south of England but increases in the Midlands and the north. In Leicestershire I have seen an oakwood on heavy loam carpeted with it—a sea of misty blue with here and there a brighter patch lit by the July sun.

In contrast, the buttercup known as goldilocks (*Ranunculus auricomus*) is essentially a plant of small compact colonies which seem to remain for many years within their chosen limits. The old naturalists often remarked on the curious irregularity of its flowers, varying from an ordinary buttercup type to a form completely lacking petals, through intermediate stages with petals variously aborted. It was thought until recently that the two extremes were two distinct varieties and that the queer-looking plants with aborted petals were hybrids between them, but it is now known that each form represents a separate apomictic microspecies (see p. 45) between which hybridisation cannot take place. The stimulus of pollination is necessary for seed-production, but actual fertilisation does not occur. The microspecies show some curious abnormalities which may throw light on inter-relationship in the buttercup family (*Ranunculaceae*); for example, the petals may be replaced by hollow nectaries, very like those found in the hellebores (see p. 63) and other genera.

A grassy ride on a hot June day is the ideal place and the ideal time for absorbing, at leisure and almost without conscious effort, the summer flora of an oakwood. Here there are light and room enough for the development of considerable colonies of species which cannot stand either the deeper shade of the wood or the full sunlight beyond its borders.

These rides are very often damp—especially on the heavier soils—and moisture-loving plants form the bulk of their flora. The blue turrets of bugle (*Ajuga reptans*, Pl. 2a, p. 39) grow in drifts just where the wood begins, overshadowed perhaps by great tussocks of the magnificent sedge, *Carex pendula* (Pl. II, p. 19). The ride itself is often thick with the devil's-bit scabious (*Succisa pratensis*) and with some of the smaller willow-herbs (*Epilobium* spp.). In this genus most of the species are interfertile, and hybrid swarms, like those in *Geum*, are very common. In rides the species usually concerned are *E. montanum, E. adnatum, E. obscurum, E. parviflorum*, the naturalised *E. adenocaulon* from North America and, occasionally in the southern counties and Midlands, the much rarer *E. lamyi*. These willow-herbs can be fairly easily identified after a little practice—Ross-Craig (1948—) has good pictures of the more difficult species—and when you come across the " sweet disorder " of a hybrid swarm, the best plan is to pick out the pure species first and then to examine the plants that do not quite tally with them. It will help to remember that first-generation hybrids are often tall, vigorous plants, overtopping their neighbours, and that their seed production is usually low.

Enchanter's nightshade (*Circaea lutetiana*) belongs to the same family as the willow-herbs (*Onagraceae*), but the four petals and eight stamens of *Epilobium* are reduced to two petals and two stamens. It is a very successful species, perhaps because it can spread both by seed and vegetatively by stolons. It makes large colonies in glades and rides but can stand fairly dense shade in beechwoods. The fruits are covered with hooked bristles and must be widely dispersed by sticking to passing animals (as our trousers and stockings testify). The stolons are formed round the base of the plant; each can produce a new plant which, in turn, forms its own stolons for future expansion.

I have, so far, been writing mainly about the flowers of woods in which the pedunculate oak (*Quercus robur*) is the dominant tree. Oak-woods on lighter, drier, more acid soils, dominated by the durmast oak (*Q. petraea*), are particularly well developed on the older rocks in

Plate 7a. Wood Sorrel, *Oxalis acetosella*. Suffolk, April
b. Winter Aconite, *Eranthis hiemalis*. Suffolk, February

Plate 8a. Wood Anemone, *Anemone nemorosa.* Kew, April
b. Common Violet, *Viola riviniana.* Suffolk, April

the west of Britain, on the Pennines, and on the gravels and sands of Hertfordshire (where they are mixed with hornbeam, *Carpinus betulus*); but they also occur in other areas where the soil is suitable, as on the Lower Greensand in Surrey and Kent. Many of the plants in the field layer are common to the two types of woodland, but in woods of the durmast oak the colonies are not so large and we miss the spring and early summer display that greets us on heavier, less acid soils. Brambles and bracken cover big areas in Hertfordshire; heather (*Calluna vulgaris*), bilberry (*Vaccinium myrtillus*) and other heath plants are abundant in the Pennine woods.

Two very characteristic plants of durmast oak woods are wood-sage (*Teucrium scorodonia*) and golden rod (*Solidago virgaurea*). The native species of *Teucrium* (*Labiatæ*) show a wider diversity in their distribution, appearance and habitats than do those of any other British genus of comparable size. The greenish-yellow flower of the tough, wiry wood-sage is an emblem of shallow, rather acid soils over almost the whole of the country. In contrast, *T. botrys* is confined to a few chalk downs and open fields in the south of England; its localities are closely guarded secrets, and places of pilgrimage to the few who know them. The third species, *T. scordium* (Pl. XIII, p. 134), is nearly as rare, but is a water-loving plant with rosy-purple flowers half hidden by leaves. Its headquarters in England are, or were, the East Anglian fens, with outposts in a handful of other counties, including one on damp sand-dunes in north Devon. It reappears in western Ireland and is quite common on limestone soils in the Shannon region. This curious distribution recalls that of the Fenland " sedge," *Cladium mariscus* (see p. 137). A fourth *Teucrium*, *T. chamaedrys*, native in central and southern Europe, occasionally escapes from gardens.

Solidago is primarily an American genus with one very variable species in the Old World—the golden rod (*S. virgaurea*). In Britain it is common on dry, heathy ground from the Shetlands to the Channel Islands and from sea-level to the highest mountain-tops. The species is equally widespread in Europe and it is not surprising that it was chosen over forty years ago by Prof. Turesson, in Sweden, as one of the plants for his pioneer work on ecotypes (see p. 44). He found that there were several genetically distinct forms from lowland, subalpine and alpine habitats, which remained true to type when transplanted into his experimental garden at Akarp. No doubt such

F

ecotypes exist also in Britain. The dwarf plants growing on mountain cliffs in Snowdonia are certainly very distinct, with broad leaves and large flowers, and have been described as a distinct species, *S. cambrica*.

The last plant of the durmast oakwoods I can mention is the little St. John's-wort, *Hypericum pulchrum*. Acid, neutral and alkaline soils each have their own particular woodland St. John's-wort; *H. perforatum* is the characteristic species on neutral loams; as soon as the lime content increases it is replaced—or accompanied—by *H. hirsutum*. Both are coarse, untidy plants compared with the elegant *H. pulchrum*, whose shiny, heart-shaped leaves and red-tipped buds make it, to my mind, the loveliest wild flower in Britain.

BEECHWOODS

Just as oak is the natural climax-woodland on lowland clays and loams, so is beech on the shallower soils overlying the chalk hills of south-eastern England and the oolitic limestone of the Cotswolds. In earlier times, the chalk hills—from the Isle of Wight to the East Anglian Heights—carried a more continuous woodland cover—much of it probably beech—but felling and grazing have now laid them bare over large areas. In patches, however, the beech remains—or has been re-planted—forming the familiar hangers of the steep downland slopes. On more level ground, where the soil is deeper, a second type of beechwood develops. These " plateau " beechwoods, as they are called, have a flora not unlike that of oakwoods on calcareous soils, with wood sorrel (*Oxalis acetosella*, Pl. 7a, p. 66) as one of the most characteristic species, but hangers, especially in the earlier stages of their development when the beeches stand close together, are unique.

The poor, dry soil gives up what little water and nutriment it possesses to the spreading roots of the beeches, whose dense leaf-canopy, in turn, allows only a glimmer of sunlight to filter through to the ground below. This double handicap—lack of sustenance and lack of light—prevents all but a handful of plants from existing on the floor of the wood. This handful, however, includes some exciting species.

There are two pale-brown saprophytes, the yellow bird's nest (*Monotropa hypopitys*) and the bird's nest orchid (*Neottia nidus-avis*), which flourish in these shadowy depths and which recall to mind W. S. Walker's haunting quatrain:

" Too solemn for day, too sweet for night,
Come not in darkness, come not in light;
But come in some twilight interim,
When the gloom is soft, and the light is dim."

The two bird's nests can exist in the hanger's " twilight interim " because they do not require sunlight to manufacture their food as do green plants, but obtain it ready-made from the decaying leaves in which they are rooted. They are enabled to do this by fungi, the threads or hyphae of which, in *Monotropa*, cover the outside of the roots and, in *Neottia*, occur actually in the root-tissue. These hyphae can apparently act as root-hairs, absorbing nourishment which is used by the plants. This association of fungus and flowering plant, known as mycorrhiza, probably began as a definite attack by the fungus, but has settled down to an arrangement of benefit to both parties. The phenomenon occurs in a wide variety of plants, including forest trees, heaths and orchids, and in many cases the " host " is unable to live without the fungus.

The bird's nests provide a good example of convergent evolution. Although so similar in their general appearance and mode of life, they belong to two entirely different families, the *Monotropaceae* (*Monotropa*) and the *Orchidaceae* (*Neottia*).

Two other genera of orchids practically complete the ground flora of young beech hangers—*Epipactis* and *Cephalanthera*. *C. damasonium*, the white helleborine, is a familiar sight in the less densely-shaded spots, its creamy, scarcely opened flowers looking rather like hard-boiled eggs garnished with green bracts. The greenish-purplish flowered kinds of *Epipactis* are difficult to classify satisfactorily into clear-cut species. They have been studied in detail by Dr. D. P. Young; his papers in *Watsonia* (1952) should be consulted by anyone interested in the group. .

In older hangers, some of the beeches which earlier contributed to the intense shade have died out, the canopy is more open, and in the wake of the increased light comes a much more abundant and varied ground flora. Dog's mercury and wood sanicle are frequently dominant, and ivy, deserting its more usual wall or tree, here invades the ground, carpeting considerable areas, often in heavy shade.

There are several species typical of these older hangers. As early as February, when the young green of the beech leaves is still a last year's memory, the little shrub, spurge laurel (*Daphne laureola*), one of the most characteristic of beech-wood plants, is already in flower. There is something romantic about an outpost, and this romance clings to the handful of British plants which are the lone—or nearly lone—representatives of large families whose main territory lies elsewhere. Mistletoe, for example, though admittedly having other claims to romance, appeals to a botanist more especially as the one British species of the great family *Loranthaceae*, with nearly 600 species, the majority of them tropical. The two British daphnes (*D. laureola* and *D. mezereum*) are in the same category; their family, the *Thymelaeaceae*, also about 600 species strong, attains its maximum development in Africa, though the genus *Daphne*, as gardeners know, has many lovely temperate species in Europe and Asia. But aside from their distinction as outposts, spurge laurel and mezereon have abundant interest in their own right. The clear green trumpets of *D. laureola*, clustered in the axils of evergreen leaves, represent a mixture of sexes, some of the flowers being hermaphrodite and some male, with stamens only. The former develop into black fruits which are exceedingly poisonous. In Britain spurge laurel is native as far north as Northumberland, but on the Continent it is a western and southern European species, and rather surprisingly does not reach Scandinavia. Though a characteristic plant of the older hangers, it is commonest, perhaps, in beechwoods on the deeper Oolite limestone soils of the Cotswolds.

Mezereon, with pink flowers, deciduous leaves and red fruits, is a much rarer species. Almost certainly native in a few localities on limestone, it has long been grown in gardens, and this has affected its wild occurrence in two opposite ways. In some of its stations it may have originated from gardens by bird-sown seeds; on the other hand, in others man has reversed the action of the birds and has dug up specimens for transplanting into cultivation.

As spring changes to summer two other plants typical of the beechwood ground flora come into flower, sweet woodruff (*Asperula odorata*) and wild strawberry (*Fragaria vesca*, Pl. VI*b*, p. 71). The woodruff is an ally of the downland squinancy-wort, *A. cynanchica*, and it has earned its epithet " sweet " from the presence of coumarin in the leaves—the same substance which gives the delicate scent to drying hay. These two Asperulas were among the many pairs of species investigated by Sir Edward Salisbury when he was comparing the average seed-weights of related plants growing in " open " and " closed " habitats.

Plate V. Marjoram, *Origanum vulgare.* Buxton, Derbyshire, July

Plate VIa. Centaury, *Centaurium minus*. Woodbridge, Suffolk, July

b. Wild Strawberry, *Fragaria vesca*. Derbyshire, July

He found, in general, that the latter had heavier seeds than the former. In woodruff, for example, the seeds weigh, on an average, about four times as much as those of squinancy-wort.

The wild strawberry, of whose delicious but diminutive fruits a prodigious quantity is essential for a worth-while helping, is not an ancestor of the garden strawberry. This arose in France about the middle of the eighteenth century as a cross between *Fragaria virginiana* from North America and *F. chiloensis* from Chile, and is a classic example of the origin of a new and valuable economic plant by the hybridisation of two species which would never have met but for man's activities. The barren strawberry (*Potentilla sterilis*), which is a very common woodland and hedgerow plant, is often confused with *Fragaria vesca*. In fruit, of course, the difference is obvious, for the *Potentilla* produces only a head of small dry nutlets; but with practice the two can easily be distinguished on floral and vegetative characters. In the *Potentilla* the small white petals have obvious gaps between them, whilst in the fuller flowers of *Fragaria* the petals touch ; and the leaflets of *Potentilla* have *spreading* hairs on the back, whilst in *Fragaria* they are covered on the back with *silky, appressed* hairs. The long thin runners of *Fragaria* are also characteristic.

Steep, high-banked paths between hangers nourish a flora quite distinct from that of the hangers themselves. Often, for many yards, the banks on both sides are thickly covered with the lovely melic grass (*Melica uniflora*), leaving, perhaps, a bare patch here and there for small colonies of wall lettuce (*Mycelis muralis*), a species with a delicacy and distinction unusual among yellow-flowered composites.

ASHWOODS

On limestone soils in the south of Britain beech, as we have seen, is the characteristic woodland tree. Farther north, its place is taken by ash (*Fraxinus excelsior*). In the Derbyshire Dales, for example, there are extensive ashwoods on the Carboniferous limestone, and farther north, on the same rock, they are well developed in West Yorkshire. Ash trees are not, of course, confined to limestone areas; they are common all over the country mixed with other trees, wherever the soil is not too acid or too dry; but it is only on the northern and western limestones that they form more or less pure woods, having a characteristic flora of their own.

If a beech hanger, with its thick overhead leaf-canopy, is the vegetational equivalent of a dim, sparsely-congregationed cathedral, the much lighter shade of an ashwood, encouraging an abundant, sun-dappled population of herbs in the field-layer, reminds us more of a modern, well-lit schoolroom, crowded with jostling children.

Many of the common ashwood plants, such as dog's mercury, lesser celandine, ground ivy and moschatel, are widespread in other types of wood, but some are especially characteristic, if not exclusive. Where the soil is dry, lily-of-the-valley (*Convallaria majalis*) may form large colonies. The history of this lovely plant in Britain is something of a mystery. William Turner, writing in 1548, had never seen it outside a garden, whereas it is now widespread, though not common, over much of England and southern Scotland. It is difficult to believe that there has really been this spectacular increase in four hundred years, and it would be worth while to examine thoroughly its printed records in Floras since Turner's day for further light on its history.

Another dry ashwood plant is the bloody cranesbill (*Geranium sanguineum*), which competes with *G. pratense* (Pl. 14, p. 116) as our handsomest native geranium. It grows often in large patches where the shade is sparse and the limestone rocks are exposed. The crimson flowers are rivalled later in the year by the autumn colouring of the much-divided leaves.

I have already mentioned (p. 68) some of the woodland St. John's-worts. There is still another, and rarer, species, *Hypericum montanum*, which enjoys the semi-shade of limestone wood-edges and bushy slopes. It is one of the finest of our dozen or so native St. John's-worts—marked off from the rest by its pale-yellow fragrant flowers in dense heads, and broad leaves with black dots along their edges.

The cranesbill and the St. John's-wort are by no means confined to ashwoods—they are plants of dry, limy, gravelly or sandy soils generally—but there is one ashwood plant *par excellence*, herb Christopher or baneberry (*Actaea spicata*). It is native only in Lancashire, Yorkshire and Westmorland, and is by no means common even in these counties. Like so many of the buttercup family (*Ranunculaceae*), it is poisonous, but has also been used medicinally.

On damp soils, the ashwood flora contains a number of species that do not thrive in the drier woods we have so far been considering. Ramsons or wild garlic (*Allium ursinum*), growing in crowded masses, is one of the commonest. It is the only British *Allium* with broad, ovate-lanceolate leaves. These resemble lily-of-the-valley, but are easy to identify, even without flowers, by their garlic smell—described, with masterly understatement, by J. W. White in his *Bristol Flora* as " an unsatisfactory perfume "—which is insistent and unforgettable. The pure white flowers are a compensation, especially when mixed, as one sometimes sees them, with bluebells and red campion—making a tricolour pattern on the woodland floor.

Like some other bulbs, ramsons forms two kinds of root; one in the autumn, consisting of thin fibres growing out horizontally and absorbing nourishment for the next growing season, and a second in April—stout, fleshy roots, which grow downwards and are contractile, pulling the bulb into the soil. These spring roots shrink eventually to about a third of their original length and cause a mature bulb to be buried about five inches below the surface of the ground.

There is only one common yellow labiate in Britain; this makes the yellow dead-nettle (*Galeobdolon luteum*, Pl. 5, p. 50), another frequent species of dampish ashwoods, easy to recognise. In books it is usually credited with two additional English names, yellow archangel and weasel-snout, and there are further local variants. I do not think that the distribution, in both space and time, of the common names of any British plant has been really thoroughly traced and recorded on a country-wide scale, and *Galeobdolon* would be a good subject for such an investigation. I have no doubt that many interesting points, in both botany and folklore, would be brought to light in the process. Anybody who travels widely in the course of their work could tackle one or two names in this way, provided they possess persistence, a scientific mind, and the art of extracting information over a glass of beer.

The commonest woodland thistle in Britain is the tall marsh thistle, *Cirsium palustre*—sometimes reaching nearly six feet in rich soil. In damp ashwoods another thistle, *C. heterophyllum*, comes into the picture. It is allied to *C. dissectum*, of damp meadows, and like it, is practically spineless; but it is a much handsomer plant with a pure white felty covering on the undersides of the leaves. The melancholy thistle, as it is called, is a good example of a northern European species with a corresponding distribution in Britain, where its southernmost stations are in Wales. In Scotland it is widespread on open pastures and hill slopes, reaching over 3,000 feet on Lochnagar.

BIRCHWOODS AND PINEWOODS

An hour's motor drive from the oakwoods of the Sussex Weald, over the crest of the North Downs, carries one into a land of sandy, acid soils with scattered birch trees golden in October, and Scots pines in dark drifts against the purple heather. These Surrey heaths and commons constitute the largest area in southern Britain of birch- and pinewood, and it is not until the borders of Scotland, 500 miles to the north, that this type of woodland is found again on a comparable scale.

There are two species of birch tree in Britain, *Betula pendula* and *B. pubescens*, and both are native over the whole country, but the great majority of the Scots pine (*Pinus sylvestris*) in England is certainly planted (see p. 54).

In general the field-layer in open birch- and pinewoods is made up of plants common also on adjoining woods and heaths, such as bracken, heather and bilberry; under closely-growing pines, however, the shade is very dense and a carpet of needles is practically the only covering to the floor of the wood.

The English pinewoods cannot boast any plants peculiar to themselves—perhaps because they are of such recent origin—but in Scotland there are a few species that are found only, or most commonly, among pines. One of these, like the two bird's nests of beech hangers, is a brown saprophyte—the coral-root orchid (*Corallorhiza trifida*). Its underground rhizomes are white and fleshy, not unlike pieces of coral, and are associated with a mycorrhizal fungus similar to that of the bird's nests (see p. 69). There is another pinewood orchid, *Goodyera repens*, which is locally abundant, but this has ordinary green leaves, and creamy-white flowers. It is one of the few British orchids with an extensively creeping, matted rootstock. Both species sometimes desert the pines for sheltered spots in old sand-dunes.

If a prize were to be awarded for the " best " closely allied group of native British plants—based on beauty, rarity and scientific interest —the five wintergreens would be very strong candidates. All five are white-flowered, evergreen herbs (though more or less woody at the base) with their main area of distribution in the north, and one (*Moneses uniflora*), the loveliest and rarest, is confined to a few pinewoods in Scotland. It has large solitary flowers, whereas in the other four the flowers are smaller and there are several on a stem. The wintergreens are closely allied to the heaths (*Ericaceae*), but are now generally separated into a distinct family, the *Pyrolaceae*. The commonest is *P. minor*, very local in woods in southern England but increasing northwards. *P. media* is very similar, but the style is longer than the stamens, whereas in *P. minor* they are equal; it is much rarer, particularly in the south. *Ramischia secunda*, another northern species, but of more open ground, is well marked by its one-sided spike of flowers. Lastly, *Pyrola rotundifolia*, with a long and conspicuously bent style, is especially interesting because, although normally a plant of moist woods, it occurs abundantly in a slightly different form (subsp. *maritima*) on the sand-dunes of the Lancashire coast near Southport.

Two other Scottish pinewood plants must complete our list. Chickweed wintergreen (*Trientalis europaea*) provides first-class ammunition against those English names, many of them copied from book to book but rarely heard on human lips, that are as uninformative as they are confusing. It belongs to the primrose family and is no relation either to the chickweeds (*Stellaria*) or to the wintergreens (*Pyrola*); and, incidentally, to add to the muddle, oil of wintergreen (Methyl salicylate) is not derived from *Pyrola*, but from *Betula lenta* or *Gaultheria procumbens*. *Trientalis* is a most attractive little plant with white flowers and the unusual number of 5-9 petals. It does not grow south of Lancashire, but in the north it is fairly common in pine- and birchwoods and on more open moors.

Linnaea borealis is without doubt the gem of the Scottish pinewood flora. It reminds us of another controversial point in plant nomenclature. To what extent should human beings be commemorated in the names of plants? *Linnaea* is, of course, beyond criticism. The great Linnaeus himself chose this lovely little shrub to perpetuate his name and asked his friend Gronovius to carry out the christening. No one could object to this, nor to other botanists and collectors being similarly honoured, but it is less certain whether it is really upholding the dignity of the science to immortalise non-botanical relatives and friends, or even men famous in other spheres of activity, by thus bestowing eponymous renown upon them.

However this may be, the delicate pink, twin trumpets of *Linnaea*, scattered over its little hummock of creeping stems and threepenny-bit leaves, provides one of the most satisfying sights in the whole British flora. It is by no means as common in north Britain as it is in Linnaeus' native Sweden, but there are quite a number of pinewoods in central and eastern Scotland where it can be found.

ALDERWOOD AND CARR

The term " carr " will be strange to most readers. It is a Middle English word, derived from Old Norse, and has had various meanings during its history. It has been revived by ecologists to designate a damp wood or copse of the type common in East Anglia.

The dominant tree in carr is the alder (*Alnus glutinosa*), mixed with ash and birch and a number of shrubs, such as guelder rose (*Viburnum opulus*) and the two buckthorns (*Rhamnus cathartica* and *Frangula alnus*). Analysis of " fossil " pollen (see p. 49) shows clearly that alder, like Scots pine, was once very much commoner in Britain than it is to-day. When, after the Boreal period, the climate became damper and milder, alder largely replaced Scots pine and birch on wet ground, as did oak where the soil was drier (see p. 53). Extensive drainage by man during the last 2,000 years has destroyed much of this old alderwood, and now, apart from small local colonies by rivers and lakes, it is extensively developed only in a few places such as the Broads, and the damp valleys of Breckland.

Tansley, in the *New Naturalist Journal*, 1948, gives a vivid description of the fen carrs, where " the air is constantly damp, so that the trunks and branches of the trees and shrubs, sometimes up to the growing twigs . . . are apt to be covered with lichens and mosses." They represent, he adds, " a very good picture in miniature of a wet tropical virgin jungle."

The ground vegetation in these " jungles " is largely composed of plants common in wet ground generally, with an admixture of waste-ground climbers such as purple nightshade (*Solanum dulcamara*), hop (*Humulus lupulus*) and large bindweed (*Calystegia sepium*) which ape, on a small scale, the lianes of a real tropical forest.

Among the damp-loving species, the purple loosestrife (*Lythrum salicaria*) is one of the commonest—and most intriguing on account of its three quite distinct types of flower, as against the two well-known types in primroses, thrum-eyed and pin-eyed. In *Lythrum* there are short-styled, intermediate-styled, and long-styled flowers; and three types of stamens, two of each being present in each type of flower. Short-styled flowers have long and intermediate stamens, intermediate-styled flowers have short and long stamens, and long-styled flowers have short and intermediate stamens. There are also constant differences between the three types of flower in the weight of the seeds, and in the size and colour of the pollen. Darwin worked out the pollination mechanism of these " trimorphous " *Lythrum* flowers, as he did that of the " dimorphous " flowers of the primrose, and revealed a situation of fantastic complication. The insect pollinators are bees, bumble bees and flies, and Darwin showed that full fertility results only when a flower is pollinated by pollen from stamens of a length corresponding to that of its own style; for example, a flower with an intermediate style pollinated by pollen from intermediate stamens. When the pollen comes from stamens of a different length, varying degrees of sterility occur.   Darwin called such pollinations " illegitimate," as opposed to " legitimate " pollinations between stamens and styles of corresponding lengths. A little thought will show that any given type of flower can pollinate " legitimately " either of the other two types but that self-pollination is always " illegitimate "; altogether there are eighteen different possible methods of pollination, six of them " legitimate " and twelve " illegitimate." This particular adaptation to ensure cross-pollination (see p. 33) is one of the most elaborate in the whole plant kingdom.

HEDGEROWS

A hedgerow is one of the most obvious and easily accessible examples of a " microclimate " (see p. 37). The shrubs and trees forming the hedge create a degree of shade, humidity and shelter which marks it off sharply from the meadow on one side and the roadside on the other. A hedgerow is, as it were, a poor relation of a wood, with something of the same ecological conditions, though much less extreme. As a result, the hedgerow flora is usually quite distinct from that of its immediate surroundings and more nearly akin to the plants found at the margins of neighbouring woods.

In recent years the destruction of hedgerows through mechanisation of farming has caused great concern to naturalists, for whom the hedgerow is part of a traditional country scene and a rich source of interesting and varied plant and animal life. Dr. M. D. Hooper of the Nature Conservancy has made a particular study of the botanical composition and importance of our hedgerows, and his paper to the

B.S.B.I. Conference on " The Flora of a Changing Britain " summarises his concern (Perring (ed.), 1970, pp. 58-62).

There are probably no species which are absolutely confined to hedgerows, but a considerable number are so typical of the habitat that they are thought of first and foremost as " hedgerow plants." Some of these are climbers which find in a hedge the same support that they would get from woodland shrubs, but, in addition, they enjoy more abundant light and air to develop freely.

One of the handsomest of these is the large bindweed (*Calystegia sepium*). Until recently it had not been realised that a still larger white-flowered alien bindweed occurred commonly in Britain. This is *Calystegia sylvestris*, a plant of south-east Europe introduced, originally no doubt in gardens, at an unknown date in the last century. The two species can be distinguished by the two bracteoles enclosing the calyx; these are not inflated at the base in *C. sepium*, and do not cover the sepals, whilst in *C. sylvestris* they are strongly inflated and usually completely hide the sepals. This, together with the larger size of the introduced species, makes distinction quite easy.

But that is not the end of the complications; for we now know that there are yet other bindweeds in English hedgerows, notably a striking pink-flowered plant whose country of origin is still somewhat doubtful, but which, like *C. sylvestris*, must have entered Britain as a garden plant in the last century.

Another handsome and conspicuous climber is the tufted vetch (*Vicia cracca*) which, in June and July, crowns hedgerows all over the British Isles with lovely racemes of blue flowers. A second vetch, *V. sepium*, prefers the lower levels of the hedgerow and is equally at home in grassy places and thickets. *Sepium* means " of hedges " (genitive plural of *sepes*, a hedge), but *V. cracca*, at any rate in Britain, has really the better right to be considered the " hedge vetch." *Cracca* is the name used by Pliny for a plant of doubtful identity, and was adopted by Linnaeus both as a name for a tropical genus of *Leguminosae*, and as a specific epithet for the tufted vetch.

A third climber, traveller's joy or old man's beard (*Clematis vitalba*), is, on limestone soils, perhaps the most conspicuous of all hedgerow plants, especially when it is in fruit. Each of these fruiting heads is made up of a number of achenes ending in long persistent styles feathered with silky white hairs. Walking along one of the deep lanes that cross the North Downs, the winter hedges often seem mantled

M. C. F. Proctor

Plate VII.   Maiden Pink, *Dianthus deltoides*. Cambridgeshire, July

Plate VIII. Lobelia urens. Grass heath, Hampshire, July

with snow when there has been a good fruiting year for the traveller's joy.

The name *Clematis* presents one of those awkward problems in pronunciation that develop when a classical word is so frequently used that it becomes virtually an English word. The " e " is long in Greek and the classically correct pronunciation is " cleematis " or " claymatis," but this sounds over-pedantic when the word is used in ordinary talk. The best solution is probably to make the " e " long when *Clematis* is linked with a specific epithet, e.g. *Clematis vitalba*, but to shorten it in normal horticultural conversation; never, in any circumstances, should it be pronounced " clemáytis."

All over Britain, from the extreme south to the extreme north, where there are hedgerows there are white umbellifers. These are usually called collectively hedge parsley, cow parsley, or kecks, but, in fact, a considerable number of different genera and species is involved. Like the yellow composites (see p. 186), they tend to form a confused tangle in most people's minds, though they are not really difficult to distinguish with a little patience and effort. Such difficulty as they do present is quite different, for example, from that surrounding the brambles (*Rubus*) or hawkweeds (*Hieracium*), where the numerous forms are distinguished by very small differences, needing an expert eye and long experience to detect. In the umbellifers the characters are well-marked and easily seen, and the despair they inspire in the beginner is really a form of laziness induced by their superficial similarity. Once a plunge has been made below the generalised umbelliferous surface, the individual species can be fished up without undue effort. At the end of the book (Appendix II) will be found a list of the commoner hedgerow and roadside umbellifers, giving their Latin and English names and brief notes on their distribution and time of flowering, together with a key for their identification, using easily seen characters of flowers and leaf. Three species stand out as perhaps the commonest and superficially the most similar; *Anthriscus sylvestris*, *Chaerophyllum temulum*, and *Torilis japonica*. As they mature one after the other from spring to late summer, only overlapping to a small extent, the time of flowering is a good first guide to identification. The characters that separate them, and their, unfortunately, numerous English and Latin synonyms will also be found in Appendix II.

A second family that provides a series of somewhat confusing hedgerow plants is the *Labiatae*. Excluding the smaller (less than 1 ft.)

species like self-heal (*Prunella vulgaris*), and those usually found in arable or waste ground (e.g. red dead-nettle) and in damp places (e.g. the mints) rather than in hedgerows as such, the following is a list comprising the larger labiates with pinkish, purplish or bluish flowers which are common in hedgerows:

Marjoram (*Origanum vulgare*, Pl. V, p. 70)
Common calamint (*Calamintha ascendens*)
Lesser calamint (*Calamintha nepeta*)
Wild Basil (*Clinopodium vulgare*)
Wild clary (*Salvia horminoides*)
Betony (*Stachys officinalis*, Pl. 2*b*, p. 39)
Hedge woundwort (*Stachys sylvatica*)
Black horehound (*Ballota nigra*)
Cat-mint (*Nepeta cataria*)

The characters that distinguish them can readily be found in a Flora and they are not at all difficult to recognise when you have once seen them and compared them with each other, but a word about the smell of their leaves may be useful to help to identify them when young and not in flower. In the first place the *Stachys* and the *Ballota* both have extremely unpleasant smells which, when once experienced, are never forgotten. Of the rest, *Clinopodium* is practically odourless, while the remaining six are all more or less aromatic, *Nepeta* very strongly so, *Salvia* only slightly. Smells are notoriously difficult to describe and the best advice is to crush and smell at every opportunity until the various species can be named at a sniff.

Lastly, I must mention what many would consider the most typical hedge-plant of all, the well-known lords-and-ladies or cuckoo-pint (*Arum maculatum*, Pl. 3, p. 46; see Prime, 1960). It is common all over Britain, except the extreme north; but there is a second, much rarer, species (*A. neglectum*) which is confined to the south of England and South Wales. The unspotted leaves of *neglectum* appear in the autumn, not in the spring, and there are also differences in the inflorescences of the two species. Forms of *maculatum* with unspotted leaves also occur not uncommonly, so this character alone is not sufficient for identification.

CHAPTER 6

MOORS, HEATHS AND COMMONS

(M. W.)

LARGE TRACTS of the north and west of the British Isles, where the rainfall is high and the rocks old, hard and acidic, are covered by moorland. In the lowland parts of England a rather similar vegetation is to be found on light sandy soil—the heaths and commons of Surrey, Dorset, Hampshire and Norfolk are good examples. Such areas are often covered by low-growing shrubs of the heath family (*Ericaceae*), the most widespread and familiar of which is the ling or heather (*Calluna vulgaris*).

Much of our moorland and heathland does not represent natural climax vegetation (see p. 40), but owes its formation, and often its continued existence, to the direct or indirect activities of man, of which the most important have been burning and grazing. On mountains above the tree limit, however, natural moorland occurs (see p. 114), and coastal windswept heaths may also be naturally treeless.

Thomas Hardy, in the unforgettable opening chapter of *The Return of the Native*, paints a poet's picture of November evening on " Egdon " Heath—a picture of sombre monotony in which the " rounds and hollows seemed to rise and meet the evening gloom in pure sympathy, the heath exhaling darkness as rapidly as the heavens precipitated it." This monotonous uniformity of moor and heath is a function of the vegetation that covers them, a vegetation made up of few species, but those few dominant over large areas almost to the exclusion of any others. The contrast, for example, between a list made on a square yard of Pennine heather moor and a square yard of chalk down turf is very great indeed; three or four species of flowering plants is an average for the former, whilst the latter may easily contain thirty or forty. Nevertheless there is considerable diversity between the various moorland types, and we can deal with them conveniently under the heads of the particular species which is dominant.

Thousands of acres of upland Britain, particularly in Scotland and the Pennines, are covered with heather moor, in which *Calluna* is the dominant, and indeed may be almost the only flowering plant (Pl. 11, p. 98). Most of the large moors are, or have been, kept as grouse moors, and their maintenance has been due, at least in part, to the practice of regular burning, which is followed by abundant regeneration of the heather from seed. Not only are large areas of young plants of heather obtained in this way, serving as food for grouse, but also those moorland plants unable to survive burning or to grow quickly and easily from seed are more or less eliminated. The heather on the heather moor is therefore to some extent a crop which has received special cultivation for many years in the cause of grouse-rearing; and much of the land in the Pennines, for example, which carries heather moor to-day was originally covered with oak (*Quercus petraea*) woodland, now almost totally destroyed.

A great deal of northern moorland on boggy, badly-drained slopes or plateaux is of the type known as " cotton-grass moor," dominated by the cotton-grass *Eriophorum vaginatum* (Pl. 10b, p. 83). (In Scotland, particularly in the north and west, the deer-grass, *Trichophorum caespitosum*, may be dominant in rather similar situations.)   This is especially well developed, for example, on the high moorland of the southern Pennines between Sheffield and Manchester, a region in which, incidentally, some of the earliest descriptive ecology was done by C. E. Moss before the first world war.   Moss's account of " The Vegetation of the Peak District," published in 1913, gives full descriptions of this desolate cotton-grass moorland from the botanist's standpoint, and is well worth consulting.   Actually we are dealing here with a transitional type of vegetation which might better be considered in a later chapter under the heading of " blanket bog " (see p. 133); but it must be mentioned here as no hard and fast line can be drawn between moor and bog.   Under conditions of high rainfall and cool climates, such as are characteristic of much of highland Britain, peat rapidly forms and accumulates on level ground or gentle slopes; and some type of peat bog formation may cover much of the land.   In general we might restrict our use of the terms " moor " and " heath " to include only those types of vegetation

Plate 9. Broom, *Sarothamnus scoparius*, by Highland river. Strathspey, June

a L. D. *Stamp*

b *Eric Hosking*

Plate 10a. Gorse, *Ulex europaeus,* in flower. Portmadoc Estuary, April
b. Cotton-grass, *Eriophorum vaginatum,* in fruit on a Yorkshire Pennine moor, June

growing where there is no great development of acid peat over the mineral soil; although, of course, it is quite likely that the non-botanist will use the terms in a wider sense.

In most moorland the main use of the land to-day is for sheep-grazing, and there are very large areas of poor quality permanent grassland on the better-drained hill slopes, in which the dominant grasses are the sheep's fescue (*Festuca ovina*) and species of bent (*Agrostis*). Such grassland is almost always below the natural tree-line, and, like most of the heather moor, owes its origin and persistence to clearing and grazing; if grazing diminishes in intensity it will often pass over into heather moorland. The trees have practically no chance of re-establishing themselves, partly because the few which are left may set little or no seed, and partly because the mortality in seedling trees through grazing and trampling will be so high. Within this sheep-grazed moorland pasture large areas are often covered with bracken (*Pteridium aquilinum*), whose power of spread is a serious threat to the hill farmer. This fern we must presume was originally a plant of open woodland (indeed it behaves as such in more continental climates such as that of Sweden), but it now occupies large tracts of land from which the forest has long since gone. When well established the plant may produce a canopy of fronds up to seven or eight feet in height which is able to suppress most other plants, including, of course, the pasture grasses; and spread is rapid by means of the long creeping rhizomes. Bracken is, however, sensitive to late frosts and exposure, and unable to grow in waterlogged soil, so that it often shows a sharp boundary, for example, at an exposed ridge, or around a marshy slope in the hills.

Two other types of grassland occur on somewhat damper soils on the slopes of moorland; these are dominated by the purple moor-grass (*Molinia coerulea*) and the mat-grass (*Nardus stricta*). Both these types occupy large areas in Scotland.

In lowland England, the heaths and commons have often been preserved because the light soil is relatively poor and unsuitable for cultivation; in this way many villages have an area of sandy heath which has escaped enclosure and on which the villagers have ancient rights of grazing their animals. On such heaths there is often a complex patchwork of different communities of plants due to varying types of interference—the effect of local fires, or varying intensities of grazing; and it is clear that some sort of woodland would in most cases rapidly establish itself if the interference slackened or ceased. The birch (*Betula*) is one of the quickest colonisers of such sandy heaths.

In the Breckland area of Norfolk and Suffolk the heaths are famous, not only to the naturalist for the rare and interesting plants and animals found on them, but also to the archaeologist for their richness in the remains of early human settlement. Moreover, the Breckland heaths have been the subject of detailed ecological investigations for thirty years, so that much is known, largely through the work of Dr. A. S. Watt (e.g. 1940), of their structure and relationships. It is now generally accepted that these sandy heaths are derived from a glacial deposit, originally largely of chalk and sand, from which in time the soluble chalk and other minerals have gradually been " leached " out, and a poor acid sandy soil left. These so-called " podsolised " soils of the Breck carry in part heather, in part grass-lands, most of which is a bent-fescue grass heath, and in part areas dominated by the sand sedge (*Carex arenaria*) or bracken. Grazing by rabbits is the main factor preventing the colonisation of such heaths by trees, particularly birch; and the intensity of grazing further determines, for example, whether heather or grassland prevails over a particular area.

Professor Godwin has suggested (1944), on the basis of his studies of peat stratigraphy and pollen analysis (see p. 49), what might have been the origin and the possible age of these Breckland heaths. He shows that the pollen of grasses, heaths, and herbaceous plants generally becomes very much more abundant in the peat record at the time when Neolithic man was colonising the Breckland area, and working, for example, the famous flint mines at Grime's Graves; and he infers that the heaths largely originated at this time by man's clearance of mixed oak forest.

It is interesting that the rare plants of the Breckland, for which the region has been famous since the time of John Ray, are without exception unable to grow on the strongly podsolised acid soils, so that over large areas of the Breck they are not to be seen. Many of these plants are weeds of cultivation, and all of them require a reasonably high lime content in the soil. The weeds are, of course, mostly annual plants which are very intolerant of competition; the other plants usually occur in parts of the grassland heaths where for various reasons there may be a high lime content in the upper soil. The speedwells (*Veronica*) provide no fewer than four of these rare

Breckland plants; three of these, *V. verna, V. praecox* (Pl. XXIIIa, p. 174) and *V. triphyllos,* are small annual plants of arable or at least disturbed ground, whilst the fourth, the spiked speedwell (*V. spicata,* pl. IX, p. 102), one of our most beautiful wild flowers, is a plant of chalk grassland. The spiked speedwell, like many of the Breckland rarities, is a continental species, that is, one which is much commoner in the drier continental parts of Europe than in our western, oceanic regions.

Many of the Breckland plants, common as well as rare, are tiny annual species which normally germinate in the autumn and early winter, survive the winter as young plants, and flower very early in the spring. By May or June, especially in a warm, dry season, they may already have ripened their seed and withered away. The little crucifer, *Teesdalia nudicaulis,* abundant on the grassland heaths of the Breck, is an excellent example of such a winter annual (see p. 30). This type of life-cycle is, of course, particularly suited to climates where the hot, dry summer is the unfavourable period for plant growth, as, for example, in the Mediterranean regions, or the warmer parts of continental Europe. A plant such as the speedwell *Veronica praecox* is able by means of this type of life-cycle to grow in very light sandy fields on the Breck where the crop is a suitable one (e.g. spring wheat), for it germinates in winter, and may be in flower by early April and have set seed by May, before the sown crop seriously competes with it.

Until the last war, there were very considerable areas of uncultivated Breckland; but afforestation, improved agriculture, and, less productively, aerodromes, have taken their toll, so that the remaining heaths are seriously reduced. The success of the Forestry Commission's pine plantations leaves no doubt in one's mind that the heaths *can* carry woodland, and presumably did so before the activities of Neolithic man first brought these large open stretches of country into being. It is inevitable that these " unproductive " areas should be put to some use by the community; but in the Breckland, perhaps more clearly than in many other places, one appreciates the case for an enlightened policy of nature conservation, so that the whole of this fascinating heathland shall not disappear.

In large areas of heathland, such as those of the New Forest, the drier parts may carry heather (*Calluna*) or bell-heather (*Erica cinerea*), whilst in the low-lying ground a type of wet heath may be developed in which the cross-leaved heath (*Erica tetralix*) and the tussock-forming grass *Molinia* may be abundant. Between such wet heath communities and valley bogs there is, of course, no sharp dividing line.

In the west of Britain heath may be developed on parts of the coast where the exposure to strong winds, and possibly also the salt spray, prevents the development of trees or tall shrubs. A particularly interesting example of such a coastal heath is found in Cornwall, on the Lizard Peninsula; here, on the serpentine rock, there is a very peculiar flora, including as one of the commonest species the Cornish heath (*Erica vagans*), hardly occurring elsewhere in Britain, but found in western France and to be seen in abundance on similar heaths in the northern regions of Spain. These western coastal heaths make a glorious picture in August. I have seen in north Devon purple heather and yellow gorse in full flower against a background of red sandstone cliffs, rich green fields and a bright blue sea.

Let us now look in greater detail at some of the interesting plants which are to be found on lowland heath and upland moor. A natural starting point is the heath family itself, the *Ericaceae*, most of whose members in Britain are moorland plants. *Calluna*, the common heather, we have already discussed, because of its importance in the moor and heath communities. Its English names are strangely confused; in northern England and Scotland it is always called " heather," and, at least in parts of Yorkshire, the word " ling " is used for the two common *Erica* species; whereas in the south " ling " is the common name for *Calluna*, and the Ericas are " heaths " or " heathers." This difference is the more curious because " ling " is obviously cognate with the Scandinavian word for *Calluna* (Swedish " ljung "); and Yorkshire, with its many traces of Scandinavian influence, in place-names and dialect, might have been expected to retain the Scandinavian meaning for the word. The folk-names of plants are a fascinating subject of study, and this example will illustrate the danger of assuming that the same English name refers to the same plant throughout the country (see also p. 73). We have decided to use " heather " for *Calluna* because of the ambiguity of " ling," and to use " bell-heather " for *Erica cinerea* and " cross-leaved heath " for *Erica tetralix*. In the last case, both the English and Latin names emphasise the character of " leaves four in a whorl " by which it may readily be distinguished from the bell-heather whose leaves are normally in threes.

Heather is a wide-ranging plant, occurring from the Western Mediterranean into the Arctic; but it is obviously most at home and achieves its greatest luxuriance in the North Atlantic fringe of Europe, including these islands. We do well to remember that our most " continental " climatic region—the Breckland—will support *Calluna* heath, and would therefore appear to the eye of a central European botanist to have a typical Atlantic vegetation. In east and central Europe, *Calluna*, like so many other western species, becomes a restricted plant of open woodland, and cannot grow in full exposure in the harsher climate.

This Atlantic type of distribution is shown also by the species of *Erica*—the gardener's heaths; even the two common wild ones in Britain are restricted to Western Europe, whilst the Cornish and Dorset heaths (*E. vagans* and *E. ciliaris*) and the rare Irish *E. mackayana* are striking examples of " Lusitanian " plants confined to the Atlantic fringe of Europe—the west of the British Isles, western France, Spain and Portugal. Only one of our moorland *Ericaceae* is rarer in West than in central Europe—this is the bearberry (*Arctostaphylos uva-ursi*), which occurs rarely in northern England and more commonly in Scotland. The bearberry is a dwarf shrub with a trailing habit of growth, and is found on rather dry scree slopes in moorland areas. On the Cairngorm mountains, where it is abundant in the lower heather zone, it is clear that the instability of the " terraced " soil is very important for the bearberry in preventing the heather from growing over it and smothering it (Metcalfe, 1950, p. 57).

As every gardener knows, the heath family as a whole is intolerant of lime in the soil, and it is useless, for example, to try to grow rhododendrons on chalk; and this is what one might expect of plants, many of whose natural homes are on the heaths of western Europe, where the soils are heavily leached by the high rainfall. There is one heath, however, which grows very well in gardens on a limy soil; this is *Erica carnea*, whose native home is in the limestone Alps, and it constitutes a rare exception to the general rule.

The bilberry or whortleberry (*Vaccinium myrtillus*) is another ericaceous moorland plant which, however, also grows in open woodland in the north and commonly high on mountains. Its deliciously flavoured purple-black berries are produced abundantly on many northern moors in July, and amply reward the patient gatherer. Its near relative, the cowberry (*V. vitis-idaea*), is also common in north Britain, and differs in having thick glossy evergreen leaves and red berries. These berries are also gathered in the Pennines and

can be seen on sale in Sheffield and other towns as " cranberries "
(see p. 139). These two species of *Vaccinium*, so very different in leaf
and in colour of fruit, are nevertheless sufficiently closely related to
produce, rather rarely, a most interesting sterile hybrid, which has been
called *V.* × *intermedium*. This hybrid, like many others, is a vigorous
plant which spreads vegetatively, so that a single plant may come to
cover a considerable area; a plant I saw in 1947 on the moors near
Sheffield covered over fifty square yards. It is strange that it is so
rarely recorded, for the parents grow together abundantly over
thousands of acres of moorland in northern Britain. The third British
Vaccinium is almost restricted to Scottish mountains; its name, the
bog whortleberry (*Vaccinium uliginosum*), is, at least so far as its behaviour
in Scotland is concerned, quite misleading, for it is a mountain plant
not particularly characteristic of the boggier soils.

The gorse (*Ulex*) is an Atlantic genus, and is in many parts of
Britain a well-known and beautiful feature of heath and moorland. It
and the closely allied broom (*Sarothamnus scoparius*, Pl. 9, p. 82) belong
to the pea family (*Leguminosae*), whose showy flowers are well adapted
for pollination by large insects. In the coastal mountains of north-
western Spain, where heathland occupies great areas, there are many
different kinds of gorse and broom, some of the latter (*Genista* spp.) very
beautiful; but of these we have in these islands only a small selection.

The common gorse (*Ulex europaeus*, Pl. 10a, p. 83), which flowers
principally in spring, is largely replaced in the west of Britain by a
dwarfer autumn-flowering species, *U. gallii*. Our third species, *Ulex
minor*, which also flowers in the autumn, is a smaller plant than either,
with much weaker spines, and occurs chiefly on some southern English
heaths. Gorse and broom are good examples of " switch plants ";
that is, plants whose leaves are reduced or absent, and in which the
manufacture of food-substances in sunlight is carried on in the thin,
bright green stems. A seedling gorse has compound leaves with three
leaflets not unlike those of clovers, but these leaves are lost early,
and the green stems take over their work. Such adaptations are often
found in plants growing in desert regions, or at least in climates where
there is a hot dry season; and it is reasonable to suppose that they are
associated with the necessity of preventing too much water loss through
the transpiring leaves. It is a striking thing that so many of our
western European heath plants show modifications of this or similar
types (cf. the small rolled leaves of heaths) although they grow best in

mild damp Atlantic climates. This is a puzzling subject, and one on which much has been written, but little clarification achieved!

We have already mentioned some of the commonest moorland grasses which cover acres of upland pastures. Another very common one is the wavy hair-grass (*Deschampsia flexuosa*), whose pretty delicate purple-and-silver flowering heads are so common in summer on acid heath and moor. Unlike sheep's fescue, which it closely resembles in leaf, this grass is strictly calcifuge; but since *some* suitable acid soil occurs in almost every county in Britain, so does the wavy hair-grass. In Cambridgeshire, for example, which has extremely little acid soil, both the mat-grass (*Nardus*) and the wavy hair-grass are remarkably rare; but they occur together at Gamlingay, just where the Greensand outcrops within the county boundary.

The harebell (*Campanula rotundifolia*), one of the loveliest and best-known of British wild flowers, provides a very good example of a plant commonly found on heaths which will actually grow in a remarkable variety of habitats, from chalk grassland to acid moorland, from sea-level to near the tops of the highest mountains. It seems, in fact, to demand only one condition—that it is not shaded by trees or shrubs (cf. *Lotus*, see p. 100). In contrast, its delicate little relative, the ivy-leaved bellflower (*Wahlenbergia hederacea*) is very much more exacting in its requirements. This pretty, slender, trailing plant is common on damp heaths and moors in the south-west of England, and tolerates or even prefers shady conditions. In the last century, to judge from old records, it was rare in shady and sheltered valleys in the southern Pennines, but the scarcity of recent records suggests that it has largely gone; and one might hazard the guess that, like the club-mosses (see p. 94), it is sensitive to atmospheric pollution, and the smoke of Sheffield and Manchester has proved too much for it. Another interesting south-western plant (too rare to have got itself an English name) is *Lobelia urens* (Pl. VIII, p. 79), which grows on a few heaths from Cornwall to Sussex. Many moorland plants flower rather late in the season, and this *Lobelia* particularly so; I have seen it in flower in mid-September on a south Devon heath. It is, of course, related to the familiar garden lobelia (*L. erinus*). Our only other wild species, the water lobelia (*L. dortmanna*), is a very different-looking submerged plant of upland ponds and lakes (see p. 157).

On the southern English heaths occur a number of tiny annual plants with very inconspicuous flowers, such as the chaffweed (*Centun-*

culus minimus), whose small round capsule releases its seeds by a neat split round the middle, just like its larger relatives, the pimpernels (*Anagallis*). Others include the all-seed (*Radiola linoides*), belonging to the flax family (*Linaceae*), *Cicendia filiformis* of the gentian family, and *Tillaea muscosa*, a dimunitive stonecrop whose bright red colour can often be seen on Breckland tracks in spring and summer. All these plants are commonly hardly more than an inch in height when in full flower; and mature flowering specimens of *Tillaea*, for example, may be under half an inch and resemble tiny mosses. It is a fascinating, if at first somewhat difficult, exercise to try to name a mixture of such tiny annuals in the field. Another very interesting annual plant of somewhat larger dimensions is *Illecebrum verticillatum*. This forms prostrate mats often several inches in diameter on damp heathy ground, particularly where the soil has been disturbed, and was until recently quite rare, almost confined to parts of Cornwall and the New Forest. Within the last few years, however, it has spread considerably in its New Forest stations, and extended its range into other counties. It will be interesting to see whether this spread continues.

Violets are very familiar wild flowers, but the countryman is usually content to distinguish two kinds only—the sweet violet (*Viola odorata*) with its delicious fragrance, and " the rest," which he would probably lump together as " dog violets." Yet the different kinds of violets deserve much closer acquaintance than this, for they are not only all beautiful in themselves, but possess a very considerable interest for the botanist in a number of ways. We will consider some of these. As is so often the case where we have a group of allied species of one genus in Britain, one of them is a common, variable and widespread plant, whilst the others are more uniform and more restricted in their occurrence. I would guess that nine out of ten wild violets found in the British Isles would belong to the single species *Viola riviniana*. This grows commonly on acid heath and moorland, even high on mountains, in a small form (subsp. *minor*) which remains dwarf in cultivation and looks rather different from the common woodland and hedgerow plant of lowland England (which is subsp. *nemorosa*); but all intermediates between the two extremes occur in intermediate types of habitats. On much English heathland, as for example in Surrey, the small *Viola riviniana*, which almost always has lilac-purplish flowers, is sometimes accompanied by another kind of violet with flowers of a rich, true blue, and narrower, fleshier leaves;

this is the botanist's " true dog violet," *Viola canina*. It may, incidentally, seem queer that the true *Viola canina* should not be the common dog violet in Britain; but we must remember that Linnaeus, who named *Viola canina*, was a Swede, and in most parts of Europe, including Sweden, this violet is common. Indeed, we find that in the Breckland, our most " continental " region, the heaths are in fact too exposed and dry in summer for *Viola riviniana*, and *Viola canina* takes its place there; although, of course, wherever there is woodland on the Breck, *V. riviniana* is likely to occur.

This picture is fairly simple; but in the south-west of England, we find a third violet on the heaths; this is *Viola lactea*, with pretty, milky-blue flowers and thick leaves even narrower in outline than the heath forms of *Viola canina*. On some heaths, as in parts of the New Forest, *Viola lactea* may occur alone, but in many places it is accompanied by one or both of the other species. When all three are present, a confusing mixture is usually found, for it seems that they are all capable of crossing with each other. All violets produce, in addition to the normal open flowers, a later crop of so-called " cleistogamous " flowers (see p. 34), and so abundant seed is set. This may be looked upon as the plant's safeguard to ensure that some seed is set, even if the open cross-pollination fails. Thus all normal plants bear capsules freely in late summer; but most of the hybrids between the different species are sterile, and quite unable to form capsules, even from the cleistogamous flowers, and they are therefore easily recognised by the complete absence of fruits. Such sterile hybrid plants are not at all uncommon wherever *V. riviniana* and *V. canina* grow mixed or in close proximity to each other; they are often very vigorous, and bear very many open and cleistogamous flowers. *V. lactea* and *V. canina*, on the other hand, seem to be able to produce at least partially fertile crosses; and therefore in a mixed population of three species, one has not only to pick out the two sterile hybrids with *V. riviniana*, but also a whole range of forms intermediate between *V. lactea* and *V. canina*! Such natural hybrid populations are fascinating to study, not only for the preliminary exercise of " sorting them out," but also to observe over a period of years the change in proportion of the different species and hybrids, and to learn something of the conditions under which such mixed populations exist. One curious feature which would repay investigation, for example, is that the sterile hybrids *V. canina* x *V. riviniana* and *V. lactea* x *V. riviniana* occur in the vicinity of their parents in large patches in several places on artificial or disturbed soil, such as heathy railway banks subject to burning, and one has the feeling that vegetative vigour is here of great advantage to them. Yet on some Hampshire heaths where there is certainly disturbed soil I have seen mixed populations of pure *V. lactea* and *V. riviniana* with no trace of hybrids.

Unfortunately there has been much confusion in British botany between the different violet species, and as a result it is often not clear exactly to which species a given reference applies. This is a pity, as the violets provide, as I have tried to show, abundant material for interesting field study on problems of species-hybrids, ecotypes, ecological preferences of allied species, and the like. The following short conspectus, based mainly on habitat, may help the beginner who is rather confused by out-of-date and somewhat conflicting accounts in our older Floras.

Violets growing in woodland or shady hedgerows :
  V. riviniana subsp. *nemorosa* : spur of flower not darker in colour than petals; sepal-appendages quite conspicuous in fruit.
  V. reichenbachiana : spur of flower obviously darker than petals; sepal-appendages remaining very small in fruit. Flowers in March and April, two to three weeks earlier than *V. riviniana* in the same place.
  Partially fertile hybrids between these two species often occur in woods where they grow together.
  (*V. hirta*, the hairy violet, and *V. odorata*, the sweet violet, also occur in woods.)

Violets growing on heath or moorland :
  V. riviniana subsp. *minor* : flower lilac, spur pale; leaves rounded in outline.
  V. canina var. *ericetorum* : flower rich clear blue, spur yellowish; leaves thick in texture, more or less triangular in outline.
  V. lactea : flower very pale milky blue; leaves thick, narrow-triangular, not cordate at base.

Violets growing in chalk or limestone grassland :
  V. hirta : flower scentless, usually pale slaty blue ; leaves triangular, usually obviously hairy.
  (*V. canina* occurs rarely.)

Violets growing on sand-dunes:
  V. canina is locally abundant on some dunes, as at Berrow, in Somerset.

Violets growing in peaty fens:
  V. canina subsp. *montana* and other forms: flowers blue; plant usually tall, with large stipules.
  V. stagnina: flowers very pale milky blue, almost white; leaves thin in texture, narrow-triangular in outline, hardly cordate; plant with thin creeping underground stems.
  These fen violets are now rare because of the drainage and cultivation of the fens.

Violets growing in acid bogs, or acid marshy places in woods, often with bog-moss (Sphagnum):
  V. palustris: flower pale lilac, very beautifully veined; leaves blunt, almost circular in outline.

There is also *Viola rupestris*, until 1960 thought to be confined to a single locality in Upper Teesdale, but now known to occur in two other limestone habitats in northern England. This rare " continental " violet should be looked for elsewhere; the British plants differ most obviously from *V. riviniana* subsp. *minor* in having dense short pubscence on the flower-stalk, petiole, and capsule.

Most of our British ferns are lovers of damp shady places, but a few are to be found in other types of habitat. The bracken we have already mentioned (p. 83); it is, of course, by far the most abundant fern in Britain, and the only fern which ever comes to be dominant over any large area in this country. It is well adapted for growth on dry, heathy ground; the underground rhizomes serve both for food storage and as an effective means of vegetative spread, and its large annual fronds when mature are quite tough in texture and do not easily wilt. The only other fern which is at all abundant on acid heath and moorland is the hard fern (*Blechnum spicant*), whose neat rosettes of tough pinnate fronds are a familiar sight on northern moors. This fern is interesting in that it bears its spores, not on the back of the ordinary fronds as in bracken and most ferns, but on specially-produced " fertile fronds " which look very different from the sterile ones. That curious fern, the moonwort (*Botrychium lunaria*), occurs also on heaths, but usually only where there is some lime in the upper layers of the soil and typical acid heath plants are absent. In the moonwort and its relative, the adder's tongue (*Ophioglossum vulgatum*), there is a single frond which bears a fertile and a sterile branch. The moonwort is not rare on Breckland heaths, and would undoubtedly be much commoner if the rabbits, which abound there, were not so fond of it; for in Watt's experimental rabbit-proof enclosures on grass heath with a high lime content in the top soil, it is quite common, though often difficult to find in the surrounding rabbit-grazed land.

The club-mosses (*Lycopodium* spp.) are another group of plants whose homes in Britain are mostly on mountains and moorland; they are remarkable plants, more primitive than the flowering plants, for their giant ancestors flourished in the great swamp-forests which made our coal, many millions of years before the flowering plants appeared on the scene (see p. 47). Now, in a vegetation-world dominated by the flowering plants, the surviving representatives of the club-mosses are few in number and generally small in stature. This is also true of the horsetails, *Equisetum* (Pl. 17b, p. 146). On mountains the upright fir club-moss (*Lycopodium selago*) is quite common; it spreads largely by means of little bulbils which are produced in the leaf-axils, readily detached, and rooting easily. The other club-mosses are creeping plants; the stag's horn club-moss (*L. clavatum*) is a very decorative plant which ought to be much commoner on moors than it now is, but has suffered a good deal from the attentions of its admirers, and, being very easily uprooted in lengths of several feet, can all too soon be destroyed. The alpine club-moss (*L. alpinum*) is, as its name suggests, more strictly confined to the higher parts of mountains, usually above 2,000 feet, where it is not, however, uncommon. The remaining species, *L. annotinum* and *L. inundatum*, are local or rare. The former resembles *L. clavatum*, but lacks the silvery hair-points to its leaves and is generally less decorative; it is rare, except on some Scottish mountain ranges. *L. inundatum* is well-named, for it is the only species which grows in *wet* heathy ground; it generally occurs in the lowlands, but is nowhere common. Club-mosses, particularly *L. selago* and *L. clavatum*, have been recorded from many English counties where they were growing on lowland heaths, but they have unfortunately disappeared from most of these. In part this disappearance is probably simply due to thoughtless picking; in part it is certainly due to destruction or change of their heathland habitats, and in part one might guess that it is due to smoke pollution of the atmosphere, for it is difficult to grow these plants in smoky town atmospheres. But it has also been suggested for these and other similar northern plants, e.g. the crowberry (see p. 119) and the mountain everlasting (*Antennaria dioica*), that their gradual disappearance from southern English stations over the last hundred years may reflect a general slight improvement in our climate. Certainly there is much evidence—for example, in the retreat of glaciers in Scandinavia and ice-caps in Iceland—that north-west Europe has been gradually " warming-up " recently, so that the suggestion is not entirely fanciful.

We have mentioned some of the more characteristic plants of heath and moor, and have seen that they fall mainly into two types, those which are " Atlantic " heath plants of western Europe, such as the heaths themselves, the gorse and *Lobelia urens*, and those northern plants which are at home on mountains but also occur on suitable lowland heath, such as the club-mosses, the crowberry and the bilberry. These are some of the plants in which visiting botanists from central Europe would always be particularly interested, for the " North Atlantic heath " is nowhere better developed than in the British Isles, with its cool but reasonably mild rainy climate, and its abundance of leached acid soils. Heath, moor and bog cover a large part of the surface of our islands, and present a picture which we ourselves are apt to take for granted, but which is startingly new to many European visitors, to whom even the common heather may be a rare and interesting plant.

CHAPTER 7

CHALK DOWNS AND LIMESTONE UPLANDS
(M. W.)

THE SCENERY which many people would claim as most typically English is that of our chalk downland, so familiar to thousands of Londoners at places such as Box Hill in Surrey. The short, springy turf of the sheep-grazed down with its bright pattern of flowers has attracted generations of nature-lovers, and provided absorbing problems for archaeologist and naturalist alike.

The English downland is conveniently divisible into four main regions, the North Downs of Kent and Surrey, the broader South Downs of Sussex, Hampshire and the Isle of Wight, the Chilterns, and the downs of Wiltshire, Dorset, Hampshire and Berkshire, in the centre of which lies Salisbury Plain. Excellent little maps are given of these areas in Lousley's book in this series (1950, pp. 55, 70, 80). What is the history of these broad acres of chalk grassland, until very recent years devoted almost exclusively to sheep-rearing? This is one of the most fascinating problems of British vegetation. We know that, in many places on the chalk, woodland occurs, and in some areas such as the Chilterns, there are very considerable beech forests; and we also know that ungrazed chalk grassland will soon be invaded by bushes, and eventually by woodland trees. It looks, then, as if the grassland of to-day is stable only because of the grazing of animals—sheep, and, of course, the ubiquitous rabbit. But exactly when our downland came into being, and what sort of woodland was there before it, we must admit that as yet we do not know. With the help of evidence provided by historians, archaeologists and others, we can elucidate fragments of the history of this or that particular area; thus, we infer from the concentration of evidence of late Neolithic culture on the Wessex downs, and, to a less extent, on the South Downs, that the original forest covering these chalk hills must have been at least partly removed by 2500 B.C.; and it may well be that in these areas a good deal of the chalk grassland turf is about 4,000 years old. But in other areas, e.g. the North Downs, there was no such widespread early culture, and the origin of the open downland must be a good deal more recent. Then again it is not safe to assume that where we now see woodland it is very ancient; for in several places "lynchets," or old Celtic ploughing terraces, can be detected in what is to-day quite native-looking beechwood, and such areas must have been cultivated a long time ago. As we have seen already (Chap. 5) our woodlands have been so much used and replenished by man that it is very difficult indeed to guess from looking at them what their complicated history has been.

Chalk is a particularly soft kind of limestone, which geologists tell us was laid down in shallow seas some sixty million years ago, which is relatively recently in geological time. What we normally call limestones are harder and older rocks. Very roughly we could say that in the British Isles the chalk areas occur in east and south-east England where the young rocks are exposed; next, to the north and west, are situated the Oolitic limestones—rather soft yellowish stone so beautifully used for building in the Cotswolds, for example; and farther north and west, in a larger arc, are to be found the Carboniferous or Mountain limestones, e.g. the Mendips, Gower, the Great Orme, the Craven district of Yorkshire, and the Derbyshire Dales (see maps in Lousley, 1950). Elsewhere in north and west Britain the limestones are small in extent and most of the rocks are ancient siliceous strata. In Ireland, however, there is a great deal of Carboniferous limestone; but little is free from peat covering in that very wet climate.

Like the chalk, limestone hills or uplands carry a great deal of sheep-grazed turf, in which many of the characteristic downland plants are to be found. As all the limestones, even the relatively soft Oolites, are much harder than the chalk, steep cliffs and " edges " occur frequently, and some of them, such as Cheddar Gorge and Malham Cove, are most spectacular and deservedly famous. Soft contours, on the other hand, are characteristic of the chalk, on which really precipitous cliffs are found (in England) only where the sea is wearing them away, for example the " White Cliffs of Dover." On these steep cliffs and screes of limestone regions it has long been known to

field botanists that many rare and local plants occur, some of them
restricted to a single small area in Britain; and largely because of
this, areas such as Cheddar and the Avon Gorge have been botanically
famous for centuries. Thus, Thomas Johnson (see p. 8), describes
a visit with a party of fellow-botanists to the already famous Avon
Gorge at Bristol in 1634 (Johnson, 1634; see also Raven, 1947, p.
287); it is amusing to find that business was mixed with pleasure, for
Johnson seems to have been staying in Bath attending a wealthy lady
patient of his! We shall mention one or two of these localities in
more detail later.

Chalk and limestone grassland is most commonly composed of a
close-cropped turf of the sheep's fescue (*Festuca ovina*). This grass is
by no means confined to such soils in Britain, however; it can, indeed,
make a good claim to be one of our most abundant plants, for it
shares dominance with species of bent (*Agrostis*) over enormous areas
of sheep-walk on the acid soils of northern Britain (see p. 83). But
other grasses which occur on chalk and limestone with the fescue are
much more restricted to these soils. A good example is the rather
beautiful silvery-spiked *Koeleria gracilis*, tufts of which are common in
many places on dry chalk or limestone grassland; and other common
calcicolous grasses are *Helictotrichon pratense*, whose stiff leaves are
covered with a whitish " bloom " which rubs off irregularly and looks
like chalk; and the upright brome (*Zerna erecta,*) whose tall flowering
heads are conspicuous in summer on many rough chalk banks.
Quaking grass (*Briza media*), well known to most country children, is
an interesting example of a plant which grows in several rather distinct sorts of habitats; it seems just as at home on a dry chalk hillside
as in a rich fen, or even an upland marsh. Such plants we suspect
to consist of several different ecotypes (i.e. genetically different
forms, each especially well suited to a particular type of habitat: see
p. 44).

It is a mistake to think that the grasses are difficult to learn. There
are not more than twenty *really* common ones in Britain, and most of
these can be distinguished not only by the appearance of their flowering
stems, but also by their leaves and basal portions. To be able to
use the last-named characters is, of course, most important for the
agricultural adviser and the ecologist,who are not often provided with
flowering shoots to aid them in their identifications. Several useful
" keys " to the vegetative parts of grasses exist; there is, for example,

Plate 11. Heather, *Calluna vulgaris*, and Scots Pine, *Pinus sylvestris*, by moorland stream. Strathspey, September

a
John Barlee

b
A. Jackson

Plate 12a. Thyme, *Thymus drucei*
  b. Bird's-foot Trefoil, *Lotus corniculatus.* Cheshire, July

a reasonably full one in Hubbard (1954) and a simpler one, omitting many rare or local species, in McLean and Cook (1946).

Although the typical short downland turf is most commonly made of sheep's fescue, chalk grassland is often dominated, where grazing is not so intensive, by taller-growing grasses, of which the commonest are upright brome and heath false-brome (*Brachypodium pinnatum*). The latter has spread a great deal in many places recently where sheep-grazing has declined, and its large yellowish patches are often quite conspicuous on the downs. Such a patch, if watched over a number of years, is seen to be extending its boundaries by outward growth of the short rhizomes from an initial centre; the relative unpalatability of the grass to rabbits means that it has every advantage over other grasses in a rabbit-infested area. The recent spread of this grass emphasises how much the composition of our downland vegetation depends upon a delicate relationship with the grazing animals, and how quickly a general change in the balance can be reflected in the vegetation. The related *B. sylvaticum* is a soft and hairy-leaved grass of woodland and hedgerow, common throughout the British Isles.

As we shall see in Chapter 9, sedges, rather than grasses, are characteristic of bog and fen vegetation, and in fact very few sedges occur in chalk grassland. Two only are at all common; these are the early-flowering *Carex caryophyllea*, a small plant with pure green leaves, and the taller, later-flowering " carnation grass " (*C. flacca*), which is easily recognised by its bluish-grey leaves. This latter, like the quaking grass, seems to grow equally well on dry chalk or in wet fen, but it is rather strictly calcicolous and avoids acid soils completely. The other sedges of chalk grassland are rare or local, and one is particularly interesting. This is *Carex ericetorum*, long known to botanists in East Anglia, where its headquarters are on the Breckland grassland —not strictly chalk, but rather chalky sand (see p. 84). For many years this sedge was not known outside East Anglia; but during the last war (1943) it was discovered (by E. C. Wallace) on Magnesian limestone in Yorkshire; and it seems that, once botanists had accepted the possibility of it occurring on northern limestone, it began to be recorded from several northern counties, including upper Teesdale, where it is growing at a considerable altitude with *Viola rupestris* (see p. 127). Although it is admittedly not a conspicuous plant, there is nothing " critical " about this sedge, which anyone can recognise in flower and fruit; and it is quite impossible to believe that the plant has recently arrived in these northern localities. We are forced to conclude that here is another case, in this so thoroughly botanised island of ours, where quite unexpected facts of distribution remain to be discovered. The " unexpected " is important; for it seems likely that one of the main reasons why botanists had failed to record this sedge from the north for so long was simply that they did not expect it to grow there, because the plant is a continental species. It is particularly interesting that the plant grows in some quantity with *Viola rupestris*, and must have been trodden on by hundreds of botanists paying their respects to this violet at what was until recently its only known British station!

Some of the commonest and most beautiful chalk grassland flowers are not confined to calcareous soils, but occur in a large variety of unshaded habitats. Such a plant is the common bird's-foot trefoil (*Lotus corniculatus*, Pl. 12*b*, p. 99), forms of which occur in practically every kind of non-wooded habitat in the British Isles, except on the tops of the highest mountains and in salt-marshes. Such a widespread species must owe its success in the plant communities found in Britain to-day to several factors, among which perhaps the most important is the ability to spread vegetatively and to survive grazing and trampling, and even to flower and fruit successfully in grazed turf. If a plant of bird's-foot trefoil is carefully dissected out of chalk turf, it will be found that there are innumerable slender stems running more or less horizontally in the thin surface soil, and capable of rooting to form a separate plant, whilst the main tap-root of the original seedling penetrates deeply into the chalk subsoil. This type of growth is admirably adapted to resist both drought and grazing, which are factors responsible for the elimination of less well-adapted species from the chalk turf. It is interesting to compare the other perennial species of *Lotus* in this country with the common one. The marsh bird's-foot trefoil (*Lotus uliginosus*), as its name says (*uligo*, moisture), is most commonly found in wet places; it is normally a taller-growing plant, though forms (ecotypes) occur in certain coastal dune marshes with more or less trailing stems ; it can grow in ordinary meadow and hedgerow only in the wet climate of the south-west. The slender bird's-foot trefoil (*L. tenuis*) is a local plant with rather specialised requirements; it is not uncommon near the sea in rough-grazed brackish meadows, and also occurs inland on calcareous soils. It is, for example, to be found on roadsides round Cambridge, mostly on the rather heavy chalky boulder-clay soils, but also (as at Cherry Hinton) on pure chalk. It differs from *L. corniculatus* most obviously in having very narrow leaflets, but also in that it seems unable to produce the slender creeping and occasionally rooting stems, by virtue of which the common bird's-foot trefoil is able to spread and flourish in heavily-grazed turf. Because of this it is undoubtedly eliminated by heavy grazing. Both *L. tenuis* and *L. uliginosus* are diploid species, whereas *L. corniculatus* is a tetraploid; (see p. 45); and as in many other cases we now know, the polyploid species is the commoner, more variable plant, existing in a number of genetically different varieties and ecotypes.

The pea family (*Leguminosae*) provides many other examples of chalk and limestone grassland plants, most of which are more strictly confined to calcareous soils than is *Lotus*. The lady's fingers (*Anthyllis vulneraria*) resembles *Lotus* in being a variable plant occurring in a variety of different habitats, including mountain rock-ledges and sea-cliffs; but inland it is restricted to calcareous, or at least basic, soils, and is a useful and conspicuous indicator of such soils. Some maritime forms of the plant, incidentally, have bright red heads of flowers and are extremely beautiful; but the common chalk and limestone plant is, of course, yellow-flowered. The horseshoe vetch (*Hippocrepis comosa*), whose dainty heads of yellow flowers are succeeded by a ring of curved, segmented pods which give the plant its name, is locally common on many English chalk downs in summer, and occurs also on limestone, although it thins out in northern England, and is absent from the Scottish Highlands and Ireland. The purple milk-vetch (*Astragalus danicus*) is one of our most interesting chalk grassland plants. It cannot be called rare in Britain, for in some places where it does occur there is a great deal of it; but it is by no means everywhere on the chalk and limestone. For example, it is absent from the whole of the North and South Downs, and the main mountain limestone areas. Not only does it occur on chalk and Oolitic limestone grassland, however, but also on many coastal sand-dunes in the north and west of Britain; and although it is absent from Ireland proper, it has an outlying station on the Aran Isles off the west coast. Such a discontinuous distribution indicates an interesting history; like *Carex ericetorum*, the purple milk-vetch is a continental plant, commoner in eastern and central Europe than in the west, and which may, we guess, have been widespread in the Late Glacial period (see p. 53).

Another very common and familiar chalk and limestone flower is the common rock-rose (*Helianthemum chamaecistus*), which occurs abundantly on many chalk downs, particularly where there is trampling or disturbance of the thin soil over the chalk and the tall grass competitors are unable to form a closed sward. The rock-roses are familiar in gardens, where many different kinds, mostly of hybrid origin, are cultivated. Even the wild plant is very variable, both in size and colouring of its flowers and also in its leaves, and in places obviously different types may be found growing side by side. The other wild perennial British rock-roses are rare or local plants in the west and south-west of Britain; the hoary rock-rose (*Helianthemum canum*) grows in several limestone regions in Teesdale, north-west England, Wales, and west Ireland (see map in Lousley, 1950); and the white rock-rose (*Helianthemum apenninum*) occurs on Brean Down in Somerset, and Berry Head in Devon. We know that *Helianthemum* was abundant in Late Glacial times; but although *H. canum* is one of the species which was present, we do not yet know whether most or all of the Late Glacial pollen finds should be referred to this. As with the purple milk-vetch, we find that the common rock-rose is almost unknown in Ireland (actually there is only one recorded station, in Donegal), and until we have more Late Glacial evidence, we can only hazard a guess as to why this should be. Two alternative theories to explain why common English species are extremely rare or absent in Ireland are, firstly, that many species were unable to cross on the land bridge from Britain to Ireland before the formation of the Irish Sea, so that they are either completely absent, or have just succeeded, through rare long-distance dispersal, in reaching and settling down in one or two scattered places; or, secondly, that species formerly abundant in Ireland in the Late Glacial period have been almost or quite eliminated in different post-glacial periods. It seems likely that both explanations are correct in particular cases. The second theory would perhaps fit the case of the rock-rose, whilst the absence of woodland plants, such as herb paris (*Paris quadrifolia*) and lily-of-the-valley (*Convallaria majalis*) is better explained by the first theory. Dr. R. L. Praeger discussed this type of problem in his volume on the natural history of Ireland (1950).

A plant very commonly associated with the rock-rose on chalk or

Plate IX. Spiked Speedwell, *Veronica spicata.* Norfolk, July

Plate X. Roseroot, *Sedum rosea*. Aberdeenshire, June.

limestone is the wild thyme (*Thymus* spp., Pl. 12*a*, p. 99), familiar to most people because of the delicious fragrance of its crushed stems and leaves, its pretty trailing habit, and heads of purplish flowers. Thyme is a very variable plant, which, as Prof. C. D. Pigott has shown (1954), can be divided into three different species, of which the rarest (*T. serpyllum*) is restricted to grassland in Breckland where it grows with other continental species. The commonest British thyme, *T. drucei*, is extremely variable, and will be found in glabrous and hairy forms with larger or smaller leaves, and ecotypes with different habits of growth. To many of these forms different names have been given in the past; but as several of the most extreme types can be grown from seed from the same wild plant, they are clearly not different species in any sense in which botanists use the term to-day. The third kind of thyme, widespread on the English chalk, is *T. pulegioides*, which usually has larger heads of flowers, and a very distinct aromatic smell which once recognised is easy to tell from the scent of the other two. These three species are nearly always recognisable in the field by a careful inspection of the hairs on the stem below the flowering heads; the differences are best explained by a cross-section diagram:

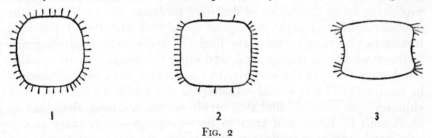

| 1 | 2 | 3 |

FIG. 2
Diagrammatic cross-sections of flowering stem of the three *Thymus* species.

1. *Thymus serpyllum*: stems not obviously angled, hairs distributed more or less evenly.
2. *T. drucei*: stems nearly square in section, hairs more or less confined to two sides.
3. *T. pulegioides*: stems nearly square in section, with two convex and two concave sides, hairs concentrated along the four ridges.

Another type of variation shown by thyme is in the sex of the flowers. Though these are normally hermaphrodite (i.e. with stamens and pistil in the same flower), many plants can be found, mixed with the others, in which all the flowers are female only. This difference is inherited, but details of the inheritance are still rather obscure.

We cannot discuss the flowers of chalk and limestone without mentioning the orchids, whose strange beauty has attracted many generations of naturalists, but as they have been described and portrayed so well in other books in this series (particularly Summerhayes, 1951), we shall be content to mention a few only. Both the fragrant orchid (*Gymnadenia conopsea*) and the pyramidal (*Anacamptis pyramidalis*) occur throughout England on suitable chalk and limestone pasture; and in places they flower in abundance in June. More locally the bee orchid (*Ophrys apifera*) may be found. It is the only British *Ophrys* which can be said not to be particularly rare and local; for the north-west of Europe is at the edge of the range of these fascinating flowers which mimic insects, and to see them in bewildering profusion and variety one needs the good fortune to visit the Western Mediterranean regions in early spring. Finally, a word about the lizard orchid (*Himantoglossum hircinum*), a fantastic rather than a beautiful plant, which has increased so strikingly both in range and abundance in England during the last hundred years, in spite of its large size and its obvious attraction for the collector and thoughtless flower-picker alike. There is good reason to believe that, with plants producing large quantities of dust-like seeds, as the orchids do, the boundaries of their areas correspond quite well with their " potential boundaries "—that is, they are likely to succeed in occupying any territory where the climate, soil, and other habitat factors are suitable for them. Thus, as the majority of our orchids are plants commoner in continental Europe and adapted on the whole to warmer, drier climates than ours, we find that orchid species are most abundant on chalk soils in Kent, and tend to become progressively rarer as one goes north and west in Britain, even where the soil is calcareous and generally suitable. The spread of the lizard orchid suggests a " softening " of our climate, for which, as we have seen (p. 95), there is other evidence. It would, incidentally, seem likely that for the same reasons, ferns, which produce dust-like spores in great abundance, manage to occupy most of the habitats available for them; but we are not justified in making such assumptions for most flowering plants, where dispersal may be a slow and erratic process depending on chance transport of heavy seeds.

The well-known family *Umbelliferae* provides several quite common chalk and limestone plants. One of these, the wild parsnip (*Pastinaca sativa*), is very easy to recognise by its coarse growth and umbels of yellow flowers; another is the wild carrot (*Daucus carota*), with its curious " half-closed " fruiting umbels, surrounded by conspicuous branched bracts. These are largely plants of roadside or waste places on calcareous soils; much the commonest umbellifer to be found on chalk grassland proper is the lesser burnet-saxifrage (*Pimpinella saxifraga*), which is a most variable plant. In particular the leaves vary enormously in the degree of dissection, so that the identification of the plant in grazed chalk grassland is a common source of difficulty to students listing the components of downland turf. Two famous chalk umbellifers are stone parsley (*Seseli libanotis*) and the great earth-nut (*Bunium bulbocastanum*). *Seseli* is not, strictly speaking, a chalk grassland plant; it is a tall stout umbellifer (not unlike wild carrot, which often grows with it but is easily distinguished by its umbels and bracts) occurring in rough grassland or banks, old earthworks, or (in Sussex) on the edge of a sea-cliff. Neither *Seseli* nor *Bunium* are at all common; and perhaps the most famous locality where both these plants grow together is at Cherry Hinton, just outside Cambridge, where roadside and hedgerow banks and old chalk-pits continue to provide them with suitable ground. Finally, two other umbellifers are particularly worthy of mention; both are very local in this country, for they have a rather southern distribution in Europe as a whole. These are the honewort (*Trinia glauca*) and the small hare's ear (*Bupleurum opacum*), a relative of the thorow-wax (*B. rotundifolium*). The honewort is abundant enough in one or two places on rather steep slopes on limestone in Somerset and Devon, and is obviously a plant which grows best in an open community rather than a closed turf. The plant is dioecious (that is, there are separate male and female plants), and if one sees a population in late summer when the male heads are withered and the females are fruiting, they look so different that one might well assume that they had really nothing to do with each other! The *Bupleurum* is one of the famous rarities of Berry Head, near Torquay, and occurs also near Eastbourne; it is an annual, extremely diminutive, even in a favourable season, and looks most unlike an ordinary umbellifer.

We have for the most part talked so far about chalk grassland and its characteristic flowers, some of which occur also on limestone. We might now consider in contrast the main features of the vegetation of the northern English limestone and some of the characteristic species found there which may be rare or absent on the warmer and drier chalk. The Craven area of West Yorkshire is one of the best known and most impressive stretches of upland limestone country in Britain. Near its southern boundary, by the famous Malham Tarn, the Field Studies Council has an excellent field centre, through the facilities of which many more naturalists have in recent years become acquainted with this remarkable countryside. The characteristic features of this limestone are the extensive areas of upland grassland, in parts with more or less continuous turf, in others, as round Ingleborough, with large areas of peculiar " limestone pavement "; the drainage is underground, and there are numerous " swallow holes," and enormous and spectacular cliffs and gorges. Hard siliceous mill-stone grit caps the limestone (with beds of shales between) on the familar flat-topped mountains such as Ingleborough, Penyghent and Fountains Fell; and the contrast between the acid, boggy vegeta-tion of the siliceous rock and the rich limestone pasture is very striking indeed. Again, as with the chalk downland, we do not know how long this upland limestone has been deforested; but the fragments of woodland still persisting at considerable altitudes in the region (as at Colt Park wood on the east side of Ingleborough) provide us with the clues we need to reconstruct the scene before extensive clearance by humans and their grazing animals. There seems little doubt that, except in the most extreme exposed ridges where there is bare limestone pavement, the region originally carried woodland largely composed of ash, at least up to 1,000 feet or so (see p. 71). Indeed, in many of the now treeless areas of limestone pavement, the deep cracks (formed by solution of the limestone along more or less rectangular "joints ") harbour saplings of ash, and a good many species, some of them, such as the baneberry (*Actaea spicata*), now quite rare, which we assume were present in some quantity in the original rather open woodland. The scrub or woodland is unable to develop again over these areas because of the intensity of sheep grazing.

If we look at the composition of the sheep-grazed limestone turf, we will find a good many species familiar on chalk down, including the often dominant sheep's fescue, the "carnation grass" (*Carex flacca*), and the wild thyme (*Thymus drucei*). But we shall also find common plants which are hardly known on the chalk. Of these one of the most striking is the grass *Sesleria coerulea*. The English name, given in some books, of " blue moor-grass " is very misleading, as the plant is strictly confined to limestone, and never grows on what most people understand by a " moor." This grass is particularly abundant on steep scree slopes and regions of open turf, where indeed it may be locally dominant; it flowers very early in the season, and in the summer period when most people visit its haunts, only the stiff erect dead fruiting heads are visible on the plants. The broad leaves have a characteristic dull bluish-green upper surface, and are borne stiffly in rather loose tussocks. *Sesleria* is a good indicator of basic soils, not only in Britain, but also throughout its range in Europe; but it seems to possess a very wide tolerance of soil moisture conditions, for it ranges, even in the British Isles, from wet mountain cliffs in Scotland to dry scree slopes and the edges of exposed pavement limestone. It does avoid hot, dry slopes, however; and it is noteworthy that, although in the north of France it occurs in a good many places on the chalk where there are cliffs or steep slopes, it is more luxuriant on the cooler north- or east-facing slopes. It would, incidentally, be interesting to compare these French habitats with superficially similar ones on the English chalk to see whether the particular combination of slope, aspect, soil and climate is not found on chalk in England; for it is surprising, on the face of it, that we have no single locality for *Sesleria* on the chalk.

Sesleria may be one of the commonest of our typically northern limestone plants, but the best-known and most conspicuous in June and July is undoubtedly the mountain pansy (*Viola lutea*). This pretty flower is thoroughly at home on the Yorkshire limestone and can be seen in great profusion in the summer; it also occurs commonly in Derbyshire on limestone grassland, but is unknown from the English chalk. Its flowers vary very greatly in size and colour; they may reach nearly an inch in length, and they may be pure yellow, as the name suggests, or purple-and-yellow, or wholly purple. Curiously enough, in different regions, different colour-forms predominate; in my experience, in both Derbyshire and Craven the yellow type is commonest, whereas in Upper Teesdale, and on the slopes of Ben Lawers, the yellow is rare, and the pure purple one abundant. Unlike the wild pansies which occur as weeds (*Viola arvensis* and *V. tricolor*, Pl. 13, p. 114), the mountain pansy is a perennial plant; it has a close relative, *V. curtisii*, also perennial, which occurs on sand-dunes, chiefly in the west of Britain.

The mountain cranesbill (*Geranium sylvaticum*) is another locally common plant of northern limestone; it is not, however, a plant of the open grassland, but rather of hedgerow, roadside, meadow or scrub. (As we have seen in Chap. 5, it is best to think of a hedgerow plant as one which in natural vegetation would occur in rather open scrub or woodland, but would avoid the dense forest.) This cranesbill almost replaces the meadow cranesbill (*G. pratense*, Pl. 14, p. 115) on high ground in the Malham district; whilst on the Oolite of the Cotswolds and Northamptonshire, meadow cranesbill is abundant.

Another plant at home on northern limestone but not on southern chalk is the lady's mantle (*Alchemilla*) in various forms (see p. 122), some in marshes and wet meadows, others on roadsides, others in grazed pasture. Ingleborough, for example, provides a perplexing mixture of these. Here, on the drier grazed limestone, particularly where the turf is sparse against a boulder, occurs the beautiful small silver-haired *A. glaucescens*, common in the Ingleborough district, but elsewhere in the British Isles known only from two areas in Scotland and one in Ireland.

As we saw earlier, steep crags and gorges, characteristic of the harder limestones but not of the chalk, provide homes for many rarities of our flora which have been known and visited by generations of botanists. In fact, it is almost true to say that any tall, steep cliff of limestone, or other basic rock, which owes its origin to natural causes will be found to have one or two—often several—rare or local plants growing on its ledges. Many examples could be given —the Cheddar pink (*Dianthus gratianopolitanus*) on the rock-ledges of Cheddar Gorge; its more widespread but equally beautiful relative, the maiden pink (*D. deltoides*, Pl. VIII, p. 78) on Arthur's Seat, Edinburgh, and several similar outlying stations; and the famous Jacob's ladder (*Polemonium coeruleum*) known to Ray, who visited it at Malham Cove in 1671 (see Raven, 1950, p. 154). Why are these rare plants restricted to one or a few limestone cliffs? In general, the explanation would seem to be this. First, these plants of rock-ledges would be killed, or at least prevented from flowering, by shade, and therefore unable to survive for long in woodland. But that would seem to be true of a great many plants common in this country in grassland, roadsides and other non-wooded habitats. There must be some reason why the rare plants do not leave the cliff-ledges and grow in these other non-shaded places. Of course, in some cases it is unfortunately true that any attempt by the plant to establish itself elsewhere than on more or less inaccessible rock-ledges is countered by the thoughtless, selfish person who digs it up for his garden or his herbarium; but this cannot be the main explanation of their restriction, for we know from the records of the first field botanists that the most showy of these rare plants were in precisely the same situations in their day as now. The explanation must surely be that nowhere except on the open cliff-ledges can these plants maintain themselves, not even in the closed grazed turf often now to be found at the top of their cliff. The *precise* reasons for this we do not know, and it is likely to be rather different in each particular case; obvious possibilities are an inability to withstand grazing or trampling, and difficulty in establishing seedlings except in loose, open soil. As yet we just have not got the type of information, derived from careful observation and experiment, which we need to help us to decide exactly what is restricting the plant in any particular case. And even when we have, an even more intriguing problem remains: how did these rare plants come to be there in the first place? In the case of the Cheddar pink, for example, its nearest known occurrences are in the Belgian Ardennes, some 400 miles from Cheddar Gorge; and almost all such strikingly local limestone rock-plants are inhabiting stations separated by many miles from the nearest European ones, and hundreds of miles from those parts of central or southern Europe where the species are reasonably common. Like the rich collection of rare mountain plants on certain calcareous places in our mountains, these limestone cliff habitats must, we feel, be interpreted as being fragments of a type of vegetation which was widespread thousands of years ago. Some of these plants, such as *Helianthemum canum* and *Polemonium*, we know were components of the Late Glacial vegetation.

It is, however, not only the herbaceous plants which we find restricted in this way to steep or otherwise permanently non-forested limestone slopes, but also certain shrubs and small trees, notably box (*Buxus sempervirens*), yew (*Taxus baccata*) and the white-beams (*Sorbus* spp.); although of course in these cases we have to try to disentangle the records of planted specimens from those which, so far as we can tell, have every claim to be considered native. The whitebeams (*Sorbus aria*, sens. lat.) are particularly interesting, for we have in the British Isles one species fairly widely distributed on the English chalk (*Sorbus aria*, sens. strict.); and then scattered on the northern and western limestones, a number of species (eight are given in the new *British Flora*) which can be distinguished on leaf shape, fruit size, shape and colour, and other characters. Some of the *Sorbus* species are very restricted, being known from only one or two crags (e.g. *S. minima* in Brecon); whilst others, such as the more northern *S. rupicola*, occur over wider areas wherever there are suitable limestone cliffs. Several of these species appear to be endemic to Britain; that is, they are not known to occur anywhere else. As with the lady's mantles (p. 122) we know that at least some of these local species are apomictic; and it is an interesting speculation as to how and when they arose. We certainly do not know; but it is a reasonable guess that, before the Boreal forest closed over large areas of the country, these and similar small, light-demanding trees and shrubs were widespread, and that the few stunted individuals we now see clinging to the face of a cliff are the remnants of a once much larger population.

We have, then, the general picture that much of the present-day chalk and limestone grassland is the result of human interference, though varying much in date of forest clearance. Very restricted areas, however, usually on steep slopes, never had a really complete forest cover, and may have retained, from the naturally treeless vegetation of the Late Glacial period ten to fifteen thousand years ago, a number of interesting rare or local species. But, of course, the rare plants in this category are merely a selection of the possible survivors, many of which have, unlike the rarities, taken advantage of man's forest destruction to extend again their areas. Such plants, we might confidently guess, are the wild thyme and the rock-rose, both of which occur in the limestone cliff refuges as well as in the close-cropped turf of the downs. This is no doubt true also of the grasses; a good example is *Helictotrichon pratense*, which appears to have a natural home on the open vegetation of limestone cliffs. Indeed, we begin to see how impossible it is, in a country such as ours whose vegetation has been so profoundly modified by man over several thousand years, to say what we mean precisely by a "natural" habitat, or a "native" plant. At its extremes, it is not so difficult; a "native" plant is one that we know, or reasonably infer, was in this country long before human civilisation developed here, and an introduced one is one brought in, intentionally or accidentally, by man. But there are all sorts of conditions between the two extremes; a truly native plant, once very restricted, may have so extended its range after human interference with the forest vegetation that it is now in every part of the country;

Plate XI. The sundew *Drosera anglica*, with Marsh St. John's-wort, *Hypericum elodes*. New Forest, July (M. C. F. *Proctor*)

Plate XII. Pale Butterwort, *Pinguicula lusitanica.* Hampshire, August

or a plant which we know was in Britain before the last ice age but which was wiped out by the ice has recently returned with human help, and has settled down like a native again; this is true of the common rhododendron (*R. ponticum*) in many parts of Britain.

Weeds of waste or cultivated ground, as we shall see in Chapter 13, are particularly difficult to classify as native or not. Most, if not quite all, weeds have, or have had, a native home somewhere in " natural " vegetation; but we do not know for certain where this was. Certain common weeds of gardens and fields, however, do occur in natural habitats, and some of them we can guess are really plants of open chalk or limestone communities which have availed themselves of the new homes provided by man. An interesting example is the parsley piert (*Aphanes*), a small annual often abundant as a weed of arable land. Actually there are two different kinds of parsley piert; one (*A. arvensis* sens. strict.), which is usually greyish or bluish-green and has small greenish flowers projecting beyond the little triangular lobes of the stipules, is common on basic soils throughout the country, whilst the other (*A. microcarpa*), usually a slenderer, purer green plant with tiny flowers more or less hidden by the finger-like stipule-lobes, seems to be confined to acid sandy soils. Now parsley piert sometimes grows on open limestone far from the nearest cultivated field. I have seen it, for example, in the Derbyshire Dales, on the rocks at Cheddar, and at about 2,000 feet on the basic rocks of the Ben Lawers range; and these plants are always *A. arvensis*. Such must be the native habitats of the plant we now think of as a weed. The other species, *A. microcarpa*, is often to be found on the bare patches of ground on sandy heaths; and here again is a likely natural habitat. The effect of cultivation has been to allow both to expand; though, as one might expect, *A. arvensis* is generally the commoner, for the farmer tries to keep his soil from going acid, if necessary by the application of lime. In the south-east of England, where much arable land is chalky soil, *A. arvensis* is the common weed; whilst in the south-west (south Devon, for example), where the soils are more commonly derived from acid rocks and the heavier rainfall works against the farmer by leaching away his applied lime, it is easy to find mixtures of the two species in arable land, the proportion we presume being largely determined by the amount of lime in the soil.

Much more could be written of chalk and limestone plants, but I have tried to convey something of the richness and interest of this type of vegetation which contrasts so strongly with the moor and bog communities occupying so large a part of Britain. I have also shown in some measure how the chalk and limestone, by providing habitats for plants more at home in central European climates than our own, have collected to themselves, through the thousands of years since the ice retreated from Britain, so many now rare and local plants which present us with many and varied problems in elucidating their history and their behaviour.

CHAPTER 8

MOUNTAINS (M. W.)

To a visitor from Switzerland, most of our mountains must seem rather dull; not only do they fail to reach a height at which their summits can carry permanent snow-caps, but their flanks do not bear the rich meadows which in early summer in the Alps are so gay with flowers. Yet to those naturalists who know the British mountains well, their flora is a constant source of pleasure and interest, with many fascinating problems and surprises.

The four main mountain areas in the British Isles, whose summits rise to 3,000 feet or more, are to be found in the north and west; namely, Snowdonia, the Lake District, the Scottish Highlands and the Irish mountains. These areas are also the regions in Britain where the oldest rocks occur; so that we have a very natural division of the British Isles between the lowland, fertile, cultivated and partly industrialised south and east, and the poorer upland north and west where the population is sparse. Most of this mountain and moorland carries a very uniform vegetation cover—which has been described in detail in a separate book in this series (Pearsall, 1950)— consisting of dwarf shrubs mostly belonging to the heather family (*Ericaceae*), or of a coarse grassland. Now there is good reason to think that the slopes of the mountains were not always so bare as they are now, but that they carried, before the spread of human cultures, a forest belt which in favourable situations might have reached up the mountain-side to a height of 2,000 feet or more. So effective, however, has been the combined felling, burning and grazing which man has caused either directly or indirectly, that there are very few places left where the lower slopes of the mountains show much trace of this original forest. Such forest remnants can be seen on the side of Causey Pike in the Vale of Newlands in the Lake District, where stunted woodland of sessile oaks (*Quercus petraea*) persists at

1,200 to 1,400 feet; and much more extensively in certain parts of the Scottish Highlands, such as the Rothiemurchus Forest of Scots pine on the northern slopes of the Cairngorm range of mountains. Anyone who has visited Scandinavian mountains will have some picture of what the Scottish Highlands must have been like before the wholesale forest destruction: in the mountain valleys and the lower slopes up to 1,500 to 2,000 feet, a pine forest; above that a zone of birch; above this a treeless heath; and finally a " tundra " zone at and around the mountain summits.

This zonation is very similar on any mountain in the temperate regions of the earth, whether in Britain, Norway, the Alps, or even New Zealand; but, of course, the particular species of plant which make up the mountain heath and the tundra may be different in the different areas. Actually we find that a surprisingly large number of the plants on British, Scandinavian and Swiss mountains are the same, and that, moreover, most of these plants are found also in the Arctic regions. Why they are on our mountains at all is a problem which has intrigued botanists for a century or more, as we shall see later.

What causes the natural " tree-line " on a mountain, above which no woodland can grow? We see the answer in the stunted, dwarf specimens of birches, for example, on the Scottish hills at 2,000 feet or so, struggling to survive at the upper limit of the forest. The mountain climate is extreme and as we climb higher it becomes more so; not only does the average temperature fall, but the exposure to strong winds increases, until on the barest, highest summits of Scottish mountains it is no unusual thing to have to struggle in the middle of summer against a roaring gale and horizontal driving sleet and hail. The winter is long, of course, and except in the most exposed places snow covers the ground for several months. In other words, as we climb a mountain we pass quickly through belts of increasingly severe climatic conditions, with their corresponding vegetation cover, very similar to those we pass through on the level if, for example, we take a train from the temperate south to the arctic north of Sweden.

The plants which will grow above the tree limit have to possess certain features in common. They are mostly low-growing shrubs or perennial herbs, which survive the winter with their buds protected by a covering of snow. Some of them, such as the moss campion (*Silene acaulis*, Pl. 16a, p. 131), are cushion plants, whose short stems

Plate 13. Wild Pansy, *Viola tricolor.* Westmorland, June

Plate 14. Meadow Cranesbill, *Geranium pratense.* Kew, June

are densely crowded together in a tight cushion offering the minimum of resistance to wind and the maximum protection to the individual shoot. Such plants are able to shelter very effectively on the lee side of rocks on exposed summits. It is indeed remarkable to see how, on the summit plateaux of the Cairngorms, for example, which are bleak, inhospitable wildernesses of granite rock and gravel, the few flowering plants such as the *Silene* and, more numerously, the small tussocky rushes, *Juncus trifidus* and *Luzula arcuata* (see p. 123), "take shelter" on the lee side of small boulders, that is, in the only places where snow can accumulate in winter and where they are therefore afforded some protection.

Another characteristic of mountain plants, which has long been of interest to the gardener, is the tendency to produce large showy flowers on dwarf stems; and although we cannot boast of the wealth of lovely alpine species which we can enjoy in Switzerland, nevertheless we can count among our native plants some of the most beautiful. To some extent the dwarf stems with short internodes can be shown to be a direct effect of growing plants in high light intensites, for light has the general effect of suppressing stem elongation; but, of course, as every rock gardener knows, the dwarf habit of growth of his alpines is largely retained in cultivation at normal lowland conditions of exposure; that is, if the plant can be grown at all, it will not look very different from the same plant growing in its native haunts.

A very few of our mountain plants are annuals, over-wintering as seeds; one of the most famous of these is the snow gentian, *Gentiana nivalis*, which is a rare plant of some Scottish mountains. In general, the scarcity of annual plants on mountains may be related to the short and uncertain summer growing season, and the consequent difficulty of ensuring the production of ripe seed every year. Indeed, many of the perennials depend far more on vegetative spread by creeping rhizomes, runners, offsets, bulbils, etc., than on seed production; and some, such as the several so-called " viviparous " arctic-alpine grasses, have replaced their flowers by vegetative buds which are shed, over-winter, and grow into a separate plant the following season (see p. 122).

It is a common experience on mountains to find that over large areas the plant cover is very uniform and rather dull, consisting of a few moorland heath and grass species, some in very great abundance; whilst most of the mountain plants are restricted to quite small areas with rather special conditions. This is largely due, as is admirably explained in Professor Pearsall's book, to the widespread formation of acid peaty soils, a process which goes on automatically, and with particular speed in regions of such high rainfall as are found in the north and west of Britain. In these podsolised soils the majority of our mountain plants are not able to grow, or at least not able to compete with the few acid-soil-loving heath and grass species. This uniformity of British mountain vegetation has undoubtedly been accentuated by the effects of grazing, particularly by sheep, and in Scotland also by deer, for mountain plants vary greatly in their capacity to tolerate grazing, and some are rather quickly eliminated. The habitats where the majority of mountain plants are to be found, therefore, are the rock-ledges on steep rock-faces, not necessarily very high on the mountain, where the soil is unstable and is continually being eroded and renewed, where little or no acid peat can accumulate, where no " closed " vegetation can develop, where there is abundant " melt " water throughout the growing season, and shelter both in summer and, beneath snow, in winter, and finally where no grazing animal can reach. The intolerance of many mountain species for acid soils is well seen in those mountain areas, such as parts of Snowdonia, where much of the rock is siliceous, giving an acid soil, but in which local veins of softer basic rocks occur. In such a region one finds remarkable contrasts in vegetation over very small areas; there are very rich assemblages of species on the basic rock-face, whilst only a few yards away on the acid rock the number of different flowering plants may be reduced to three or four. The starry saxifrage (*Saxifraga stellaris*, Pl. 16*b*, p. 131) is one of the very few arctic-alpine plants which is to be found commonly in regions of acid rock; in contrast, the purple saxifrage (*Saxifraga oppositifolia*, see p. 120) is a faithful indicator of basic rock.

Although in the British mountains to-day we have practically no permanent ice and snow, there are a good many places, particularly on the higher Scottish ranges such as Ben Nevis and the Cairngorms, where large snow-patches persist in sheltered hollows, usually facing east or north-east, until late in the summer. Around such a snow-patch there is a rather precise arrangement or zonation of vegetation; and, in fact, hollows where the snow lies late into summer may easily be recognised on Scottish mountains by the types and arrangement of the plants growing there, even if at the time there is no snow persisting.

As we have seen, *some* protective snow cover is essential for many of the plants which grow on mountains; but, of course, if the snow does not melt quickly in spring or early summer, the available growing season is too short for many plants; and so we find that the late snow-patches have their characteristic species, and many common mountain plants are excluded from them. Fig. 3 shows diagrammatically the sort of zonation which is usually seen. Ideally one might picture a

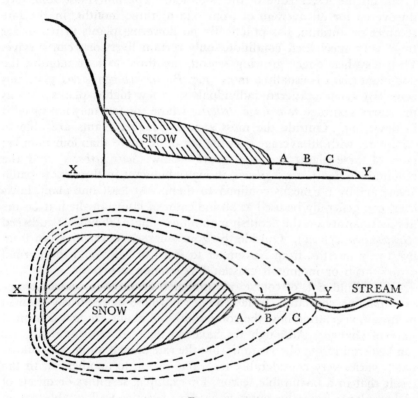

FIG. 3

Diagrammatic representation of snow-patch zonation in late summer (simplified, from snow-patches observed in the Cairngorms). (*above*) Section; (*below*) Surface view.

A. Innermost liverwort zone.
B. Moss zone, with scattered flowering plants (e.g. starry saxifrage, *Saxifraga stellaris*).
C. Zone with snow-patch flowering plants (e.g. *Salix herbacea*, *Alchemilla alpina*).

permanent snow-patch as circular and lens-shaped, situated on level ground, and surrounded by rings or zones of different vegetation, the inner zones next to the permanent snow being uncovered for only a very short period each season. In fact, of course, one never sees the zonation quite so clearly; all snow-patches are situated in depressions (usually of the type shown in Fig. 3); they are usually irregular in shape, and the drainage channels made by the snow " melt water " break up the lower edge of the zonation. The innermost zone (A), uncovered for an average of from one to three months, in the late summer or autumn, has practically no flowering plants in it; under these very specialised conditions only certain liverworts can survive. With a rather longer growing season, say three to four months, the dominant plant is usually a moss (e.g. *Polytrichum norvegicum*); in this zone (B) occur scattered individuals of a few higher plants such as the starry saxifrage (*Saxifraga stellaris*) whose rosettes may not succeed in flowering. Outside the moss zone the higher plants are able to dominate, with an average growing season of more than four months; most of these plants, e.g. the dwarf willow (*Salix herbacea*) and the mountain sedge *Carex bigelowii*, though abundant in these snow-patch areas, are by no means confined to them. At least one plant, however, can generally be used as an indicator of late snow-lie if it occurs in any quantity on the Scottish mountains—that is the dwarf cudweed (*Gnaphalium supinum*). Outside this zone the snow cover may last from five to six months, from November to April or May, and the normal dwarf shrub or mountain grassland vegetation is present.

We should not, of course, picture such an arrangement of plants as remaining constant. In fact we see nowhere more clearly than in mountain vegetation how the communities of plants are in a continual state of change. Although we have no detailed measurements, we can be certain from observations that the rate of melt of the same snow-patch varies very considerably from one season to another; with the result that in a favourable season, for example, seedlings or offsets of higher plants from the outer zone may become well established in the inner moss zone—or even, rarely, in the innermost liverwort zone—before the snows again cover them. All the zones then tend to move inwards. An unfavourable season, however, may kill or seriously affect many plants by cutting down their growing season to an exceptionally low level, and thus force the zonation outwards.

One of the most interesting features of the snow-patch plants is the adaptations they show to the peculiar conditions under which they grow. We lack the striking examples of the Alps, where the pretty Soldanellas, for example, can actually flower through several inches of as yet unmelted snow at the edge of the patch. Yet dwarf cud-weed (*Gnaphalium supinum*) and alpine lady's mantle (*Alchemilla alpina* —a common snow-patch plant, see p. 121) both show undoubted adaptation in the possession of a silvery coating of hairs, which, by preventing wetting, enable the plants to start their growth quickly on the very edge of the snow where the ground may be running with melt water. And it is also worth mentioning the truly remarkable and unexplained powers of survival of some of the lower plants of the inner zones which must be able to survive consecutive seasons of total snow cover. Mr. H. Seton Gordon has written a very interesting account of the plants of Scottish snow-patches in a letter to *Nature* (Gordon, 1950). Describing the famous " permanent " snow-bed on Braeriach, he says that, during his observations, which extend over a period of forty years, he has seen bare rock, devoid even of lichens, at the edge of the melting snow on only three occasions—the last of these occasions was in October 1949, when the snow had almost disappeared. He is certain from his observations that some of the lower plants can survive up to six years' continuous snow cover!

I have talked here in some detail about snow-patches, partly because what we know of their vegetation is so interesting, and partly in the hope that the reader with a taste for mountains may be inspired to do a little study, for example, of precise vegetational changes at the same patch from year to year, and so fill in some of the gaps in our knowledge.

Let us now look at some selected examples of British mountain plants in greater detail; the examples have been chosen to illustrate a wide range of different types.

The crowberry (*Empetrum*) is a very common plant on British mountains, wherever the rock is acid or the soil peaty; it occurs also commonly on moor and heathland, but becomes very rare in the south and east of England. Actually we have in Britain two different crow-berries, which are now generally distinguished as species; one of these, *Empetrum hermaphroditum*, is a real arctic-alpine plant, restricted to the Arctic regions and to the mountains of North Temperate Europe, and the other, *E. nigrum*, is a more lowland plant. As the name suggests, the former bears bisexual flowers, and this condition can be recognised even when the plant is in fruit, as the withered stamens persist for a long time; *E. nigrum* is normally dioecious, that is, the male and female flowers are borne on separate plants. It seems also that there are other differences, less easy to describe, by which the two plants may usually be distinguished; *E. hermaphroditum* is the stockier plant with a less straggling growth and broader leaves. Finally, it has been shown that *E. hermaphroditum* has twice as many chromosomes as *E. nigrum*. On many Scottish mountains where the crowberry occurs abundantly in the dwarf shrub zone, most of the plants above about 2,500 feet are *E. hermaphroditum*. In north Wales, however, the two species seem to be more mixed, and I have gathered both, for example, on the summits of the Glyders at more than 3,000 feet. It would be interesting to know more about their distribution on British mountains. A final point of interest shown by the crowberry is that it flowers rather remarkably early for a mountain plant—I have seen it in full flower at Easter in Snowdonia and the Lake District at a time when hardly any other mountain plant is flowering.

The mountain avens (*Dryas octopetala*) is a familiar and beautiful rock garden plant, and undoubtedly one of our finest native mountain flowers. Like the crowberry, it is a common plant in the Arctic, and occurs on many mountains in Europe. Its soil requirements, however, seem to be the opposite of those of *Empetrum*; it is a calcicole, growing only on limestone or other basic rock. It is not therefore surprising to find that *Dryas* is a local plant in the British Isles, and absent from large areas of mountain where there is no suitable soil for it. What is at first more surprising, however, is to find that on limestone in the extreme west of Ireland this arctic-alpine plant is abundant almost at sea-level. There has been much discussion of this and other peculiar distributions shown by some British plants, and we shall consider these problems later (see p. 128).

The purple saxifrage (*Saxifraga oppositifolia*) like *Dryas*, is a familiar alpine of the rock gardener, and a common Arctic plant which is found on European mountains. When not in flower, the straggling stems with opposite leaves may be mistaken at a casual glance for wild thyme (*Thymus*), but when the stems are beset with the beautiful large purple flowers in April or May, it is a glorious and unmistakable sight, amply rewarding the naturalist who visits at Easter the wet rock-faces and slopes which are its haunts in Snowdonia and elsewhere. At this time, most other mountain plants to be found with it have hardly yet started to grow. I have already mentioned that the purple saxifrage grows only on base-rich soil; and in the great corries of Braeriach and Cairngorm, for example, one can search in vain for the plant, for the rock is a coarse, acid granite; whilst on much less well-known and lower mountain rocks in the vicinity, where there is a vein of basic rock, the plant occurs abundantly. The yellow saxifrage (*S. aizoides*), another Arctic-alpine plant, also requires a base-rich water supply; but it well illustrates the kind of unsolved problem which abounds in the study of plant distribution, for it occurs quite frequently in suitable spots in Scotland and the Lake District, but is completely absent from north Wales.

The alpine lady's mantle (*Alchemilla alpina*, Pl. 15, p. 130) also occurs in the Arctic and in the Alps as well as on our mountains; but is an oceanic plant, which grows abundantly in the wetter Norwegian coastal mountains but is absent from the drier continental mountain ranges of northern Sweden. To some extent its behaviour in Britain fits this pattern quite well; for example, in the Lake District the plant is very common in the western hills—it can be seen in abundance in Honister Pass, for example—and becomes very much rarer as one goes east. However, it is unaccountably absent from Snowdonia, and in Ireland occurs only in a very few places in the mountains in the south. Again, whereas on mountains such as Ben Lawers it is very common and by no means confined to one particular kind of habitat, in the Cairngorm massif it is almost confined to late snow-patches. Now this may be explained generally by the facts that the Cairngorms are drier than the more westerly hills, and have, moreover, a very coarse and well-drained granite soil, so that only the well-watered snow-patch areas are suitable for its growth; but the plant *does* grow on outcrops of rather dry limestone rock in the Cairngorm region at quite a low level of about 1,000 feet. It seems that to some extent *Alchemilla alpina* behaves as a calcicole in dry mountain regions, whilst in regions of heavier rainfall it is more or less indifferent to soil. This type of behaviour is not unknown in other species; thus the common primrose (*Primula vulgaris*) avoids the poor soils of the gritstone and the coal measures in the southern Pennines, whilst in the rainier south-west of England it seems more indifferent to soil; although Good (1944) has shown that in Dorset it tends to avoid the chalk, which is too dry for it (see p. 60). No one has yet suggested a satisfactory explanation of this type of distribution; and again we are aware of our ignorance as to why plants grow where they do.

All the British lady's mantles (*Alchemilla* spp.) are apomictic—that is, they set seed without fertilisation (see p. 45); the pollen is usually " bad " and will not ripen properly. This state of affairs is not uncommon in the rose family, *Rosaceae*, to which *Alchemilla* belongs, and in the dandelion family, *Compositae*; both the bramble (*Rubus fruticosus*) and the dandelion (*Taraxacum*) are examples of apomictic plants. It is often associated with the occurrence of a large number of slightly differing forms, to which the name " microspecies " is sometimes given. The common lady's mantle (*Alchemilla vulgaris*) consists in Britain of about a dozen such microspecies, which differ in general size, hairiness, leaf-shape, and other characters; and in the Alps there are a great many more of these forms. Curiously enough, however, *Alchemilla alpina*, though apomictic, looks almost precisely the same whether growing in Greenland, Iceland, Scotland or Scandinavia—although in the Alps there are a number of different microspecies, of which ours is one. We saw earlier (p. 115) how many arctic-alpine plants have adopted " short cuts " of various kinds to make sure of perpetuating their kind (by offsets, bulbils, etc.); and apomixis may well be another of these devices by which the uncertainty of cross-pollination is circumvented. It must be significant that the north temperate flora has rather a large number of examples of apomixis; and the suggestion has been made that in the conditions prevailing when the ice was retreating at the end of the last ice age, those plants, such as the apomicts, which were able to produce abundant and regular seed colonised the newly uncovered land most quickly and were at the time most successful.

The common alpine lady's mantle of gardens is not *Alchemilla alpina*, but a taller, rather more decorative plant with leaflets joined at the base, known as *Alchemilla conjuncta*. This plant, unlike the vast majority of our mountain plants, grows in the Alps but not in Scandinavia or the Arctic. It is well known in Glen Clova, where it was recorded in the last century. Some botanists have suspected that it was accidentally or deliberately introduced—for it is an easy plant to cultivate.

Alchemilla faeroensis, which is rather similar to *Alchemilla alpina* but by no means identical, occurs in the Faeroes and in

Iceland, but nowhere else, and it is intriguiug to speculate how and when it arose.

Lloydia serotina is a British mountain plant with many claims to distinction. It was first described by Edward Lhuyd, Welsh naturalist of the seventeenth century (see p. 16), and named in his honour by R. A. Salisbury; it is undoubtedly Snowdonia's most famous botanical rarity, for the north Wales mountains provide its only habitats in the British Isles. It is a diminutive member of the lily family (*Liliaceae*), and like many of its relatives, possesses a bulb by means of which it survives the winter. Now plants with bulbs, corms and deep swollen rhizomes—underground storage organs—are very common in the warmer parts of Europe, particularly in the Mediterranean region, where the growing season for plants is the mild rainy winter and early spring, and the plant withers above ground and survives the hot, dry summer in the resting underground structure. In cold climates, however, these so-called " geophytes " are very few (see p. 35). The reason seems to be that in arctic (and therefore to some extent mountain) climates, the soil may be permanently frozen below the surface layers which thaw out in summer, so that plants with deep underground parts are unable to start their growth quickly in the short summer season. Thus those plants which over-winter by means of buds at or near the soil surface, that is, the tufted herbs and dwarf shrubs which, as we have seen, are the common types of mountain plants, have a great advantage over the other types in very cold climates. *Lloydia* constitutes the rare exception to this generalisation. Mr. N. Woodhead has given an interesting account of the ecology of this unique little plant (Woodhead, 1951a).

Remarkably few of our mountain plants are strictly confined to the tops of the mountains; the majority occur occasionally, some frequently, at very much lower levels. For example, the dwarf cudweed mentioned above in connection with snow-patches (see p. 118), rarely occurs below 2,000 feet, and then only as isolated weedy-looking individuals by stream-sides. Perhaps the only locally common mountain plant which is strictly confined to the exposed summits of Scottish mountains is the tiny arctic rush, *Luzula arcuata*, which is rarely, if ever, found below 3,000 feet. This plant is a true arctic member of the flora, and absent from the Alps. It resists the extreme conditions of exposure to be found on the bare summit plateaux of the highest Scottish mountains, and in similar situations in the Scandinavian arctic mountains one of its few companions would be the lovely snow buttercup, *Ranunculus glacialis*, which, alas, does not occur in the British Isles.

It is well known that several of the rarest plants in the British flora are mountain plants. For many years it had been supposed that the arctic-alpine saxifrage, *Saxifraga cernua*, occurred on a single mountain in Scotland only, but in the summer of 1949 it was found in an entirely different locality. This discovery emphasises the fact that the less famous Scottish mountains are still relatively little known botanically—even a mere catalogue of the plants growing there may not exist. Later we had even more striking proof of this by the discovery of three more arctic plants on Scottish mountains which had never previously been recorded in this country. One of these is a tiny, inconspicuous annual, *Koenigia islandica*, and perhaps it is not surprising that this plant was missed; but of the other two, *Artemisia norvegica* and *Diapensia lapponica*, the latter is a cushion plant with quite large and showy flowers—and this was first seen by an amateur ornithologist visiting mountain-tops which most botanists thought too dull to repay their attention!

Let us turn now to some examples of mountain plants which are not arctic or alpine in the restricted sense. As was explained earlier, over large areas of British mountains certain common species are abundant (the crowberry, *Empetrum*, discussed on p. 119, is a good example); in fact the commonest plants on many mountains are plants we are familiar with in the lowlands also. Thus, heather (*Calluna*) is abundant and may be locally dominant up to 3,000 feet and higher on some Scottish mountains; the bilberry (*Vaccinium myrtillus*) is locally dominant up to 3,500 feet and occurs in stunted form on and near the highest summits; whilst such grasses as the mat-grass (*Nardus*) and the tufted hair-grass (*Deschampsia caespitosa*) occur at all heights from sea-level to the highest summits and are often abundant on the mountain slopes. Other plants which never occur very abundantly may nevertheless be found at all heights on the mountains; such a plant is the common dog violet (*Viola riviniana*), dwarf forms of which occur at over 3,500 feet in Scotland; and the red campion (*Silene dioica*) is a quite common and beautiful addition to the rich flora of high rock-ledges. The rock-ledge flora is, indeed, usually composed of a very interesting mixture of lowland and arctic-alpine species, the former often represented by rather distinct-looking forms, or ecotypes (see p. 44), adapted to the special conditions of the mountain habitat. I recorded, for example, the following species, among others, growing at 2,600 feet on wet rock-ledges in Cwm Glas, Snowdon (September 1950):

Lady's mantle (*Alchemilla glabra*)
Wild angelica (*Angelica sylvestris*)
Sea pink (*Armeria maritima*)
Green spleenwort (*Asplenium viride*)
Harebell (*Campanula rotundifolia*)
Ox-eye daisy (*Chrysanthemum leucanthemum*)
Mountain sorrel (*Oxyria digyna*)
Mossy saxifrage (*Saxifraga hypnoides*)
Purple saxifrage (*Saxifraga oppositifolia*)
Starry saxifrage (*Saxifraga stellaris*)
Alpine saw-wort (*Saussurea alpina*)
Rose-root (*Sedum rosea*, Pl, X, p. 103)
Moss campion (*Silene acaulis*)
Devil's bit scabious (*Succisa pratensis*)

To some extent, the separation of these lowland plants on high mountain ledges from their main woodland populations is an artificial result of the forest destruction and over-grazing of mountain slopes. It seems likely that at least on mountains where the rock is not too acid, the well-drained slopes should carry scrub and in the wetter " seepage areas " rich " alpine meadows " of plants such as the globe-flower (*Trollius*) and the cranesbill *Geranium sylvaticum*. In these communities many lowland woodland species would naturally occur. But what we almost invariably find is that the upper edge of the forest, with almost all the scrub and alpine meadow zone, is absent, and mere fragments of these communities remain clinging to steep rocks where grazing is impossible. Some of our rarest mountain plants are undoubtedly casualties of the upper forest destruction; one example is the beautiful blue-flowered alpine sow-thistle (*Cicerbita alpina*) which is restricted to more or less inaccessible rock-ledges in the Scottish Highlands, where it escapes the attention of sheep and deer. In the Norwegian mountains this stately plant is often abundant in the upper part of the coniferous forest stretching up the mountain-side.

Other mountain plants whose relative rarity is, one strongly suspects, due in large measure to over-grazing are the smaller shrubby willows. These willows are usually very difficult to name, for they all hybridise freely, and since the sexes are separate, cross-pollination must take place if any seed is to be set. Indeed, it is no exaggeration to say that hybridisation has gone so far in the willows, particularly the arctic-alpine ones, that even the experts cannot always decide or agree as to what the parent species are. A Swedish botanist succeeded in breeding a willow with no fewer than fifteen different species in its immediate ancestry! Perhaps the finest of these native willows is *Salix lanata*, which, as its name suggests, has leaves which are covered with a thick coat of whitish woolly hair. It is a rare pleasure to see a patch of these beautiful dwarf shrubs on some steep rock slope in the Highlands, growing with globe-flower, alpine meadow-grass (*Poa alpina*) and many other mountain plants.

To return to the lowland species which have mountain forms; in a few cases, for example, that of the thyme-leaved speedwell (*Veronica serpyllifolia*), the mountain plant looks sufficiently different from the ordinary lowland form to have received a different name (*V. humifusa*). Indeed, on mountains where *Veronica serpyllifolia* does not occur on the slopes, and *V. humifusa* is confined to wet rock-ledges high on the mountain, the two plants remain remarkably distinct; but in other places all kinds of intermediate forms may be found at different heights on the mountain. It seems best, to most botanists, to treat such plants as two subspecies of a single species rather than as two separate species.

A similar example, which has generally been treated as a variety or subspecies in this country, but often called a distinct species by Scandinavian botanists, is the mountain form (subsp. *fontanum*) of the common mouse-ear chickweed (*Cerastium vulgatum*). This plant has rather larger flowers, and larger seeds, than the ordinary lowland weed. It is quite common on our mountains, where it may grow with the true arctic-alpine species *Cerastium alpinum*; on Ben Lawers the two species are commonly found growing together, and with them a sterile plant which looks like a hybrid. Two other mountain *Cerastia* occur in Britain, one, *C. arcticum*, is a rather rare plant of some Welsh and Scottish mountains and the Shetlands; while the other, *C. cerastioides*, is more widespread, and seems to be more common on siliceous than on basic mountain rocks.

A rather remarkable and familiar group of mountain plants are those which occur also by the sea, and are often absent from most types of habitat in between. Such are the sea campion (*Silene maritima*), the sea plantain (*Plantago maritima*) and the thrift or sea pink (*Armeria maritima*). Both the campion and the plantain have been studied in some detail (see pp. 176, 179), the latter particularly from the point of view of ecotypic variation; and the mountain and seashore populations of these plants do differ in heritable characters of habit and growth, time of flowering, length of stem, and many others. The explanation of these links between mountain and seashore vegetation must be sought in the history of our mountain flora, which is discussed below.

One of the most famous botanical areas in Britain is that of upper Teesdale, on the border of Yorkshire and Durham, which possesses a unique assemblage of plants, some of which occur in no other part of the British Isles. A number of these interesting species occurring in Teesdale are true arctic-alpines—for example, the rare " sedge " *Kobresia simpliciuscula*, and the tiny sandwort *Minuartia stricta*—whereas others of the " Teesdale assemblage " are not really mountain plants but rather plants characteristic of the drier steppe-like vegetation of parts of northern Continental Europe and Asia. The most famous of these plants is the shrubby cinquefoil (*Potentilla fruticosa*), familiar to gardeners in many varieties, which grows abundantly on the gravelly banks of the River Tees, and is a charming sight in flower in June and July. It occurs also on mountain screes and high rock ledges in the Lake District, in one locality at more than 2,000 feet, and so has a good claim to be considered a mountain plant in Britain. Like the mountain avens (*Dryas*), however, we find it abundant on the very peculiar limestone pavement near sea-level in County Clare, in the west of Ireland: and remains of both it and *Dryas* have been found in deposits in the south of England dating from the period after the retreat of the ice but before the main spread of forest (see p. 53). Again, the Teesdale area possesses the most famous British station for the little violet *Viola rupestris*, a plant of a very similar continental distribution. It is interesting to note that this violet has very similar isolated mountain habitats in arctic Scandinavia, separated by hundreds of miles from the main occurrences of the plant in central and south Sweden. This remarkable assemblage of rare plants has achieved fame in recent years in connection with the building of a reservoir at the head of the valley, a project which was resisted by the Botanical Society of the British Isles and the nature conservation movement generally. The new reservoir submerges part of the unique vegetation, and may profoundly affect other areas near the new water-level. The case for their conservation is given in Godwin and Walters (1967).

As was mentioned earlier, the presence of arctic-alpine plants on mountains led English naturalists, over a hundred years ago, to speculate on how they came to be growing there, and from that, on the more general problem of the significance of the facts of plant distribution in terms of the history of changes in our flora. We owe to Edward Forbes' classical address to the British Association in 1845 the first clear statement of the hypothesis that arctic-alpine plants were " left behind " on our mountains when an arctic type of vegetation followed the retreating ice-sheet northwards after the ice age (see Forbes, 1846). This general theory, which Darwin stated at length, has not been seriously upset or greatly modified during the century that has elapsed, although, as we saw in Chapter 4, it was not until recently that we had any direct evidence as to the actual plants which were growing in Britain, and elsewhere in northern Europe, at successive periods since the ice retreated. We do now know that over a large area of Britain, as the ice melted, there was a remarkable " open " vegetation on ground covered by rock debris left by the ice and freely supplied with melt-water—the glacial gravels, boulder-clays, etc. This Late Glacial period, which may have lasted for thousands of years with climatic fluctuations, must have been the time when many of the plants we have been discussing in this chapter were abundant all over the lowlands of what is now Britain (not then separated from the Continent). On the high mountains at that time glaciers and per-manent snowfields still lay. Thus the crowberry must have covered large areas where the rock debris gave an acid soil, or when sufficient acid peat had accumulated; whilst over the enormous areas of chalky boulder clay and other calcareous soils there must have been a low-growing vegetation carpet in which *Dryas* was abundant. At some stage in this arctic tundra vegetation there entered Britain from the Continent a number of plants characteristic of continental or even steppe conditions, of which the shrubby cinquefoil was certainly one, for, as mentioned above, its seeds have been identified from Late Glacial deposits in south-east England. We may also assume, though the direct evidence for this is not yet available, that some of the abundant pollen of the rock-roses (*Helianthemum* spp.) identified in

Late Glacial deposits was contributed by the now remarkably local hoary rock-rose (*H. canum*—see p. 102), and further that plants such as the Teesdale violet, and the sedge, *Carex ericetorum* (see p. 99), were quite widespread at this time in England.

Our picture, then, is that in only very few places in the British Isles have fragments or relics of this formerly widespread Late Glacial flora of 12,000-15,000 years ago been able to survive. As we saw in Chap. 4, over this open Late Glacial vegetation spread the forests, in precisely the order in which they are (or were before man's destruction) to be found as one descends a Scottish mountain. Thus first came the scattered birches, then a more complete birch scrub or woodland, then pine forest (in the Boreal period), then deciduous forest, with trees such as oak, elm, and lime with which we are familiar in lowland Britain today. Now the plants common in the Late Glacial have at least one thing in common; they are intolerant of competition, and many actually demand an open vegetation, with a good deal of bare ground. We are very ignorant of the real factors which prevent many mountain plants from growing in " closed " communities, but in many cases it seems likely that the plants simply cannot tolerate the shade cast by their taller-growing competitors. Thus, when the trees colonised the open landscape, they brought about the extinction over wide areas of this earlier flora, which might survive only in places where the trees could not grow, such as the mountains, the sea-coast, rock-faces and eroded river-banks. In this way the formerly continuous populations of plants such as the sea campion were separated into coastal and mountain forms.

There is also another factor which must have caused further restriction, particularly of open mountain communities. As the climate became wetter, in the Atlantic period, and especially in the most recent Sub-Atlantic period (which began some 2,500 years ago), the formation of acid peat became very widespread, and thick " blanket-bog " (see p. 133) spread over much hill country. Now, as we saw earlier, the vast majority of arctic-alpine plants, and also other plants common in the Late Glacial, cannot grow in acid peat, which indeed carries a poor and specialised vegetation; and so they were severely restricted even where they had escaped the woodland. The rather few places where neither woodland, scrub nor peat can form are the places where we find rich and diverse flora today; in Teesdale, for example, on the eroding bank of the river, where the calcareous morainic debris is constantly slipping away, or on bare, gravelly sugar limestone; and in County Clare, on the exposed fissured pavement limestone. It is interesting to remember, however, that the Late Glacial plant which is abundant on mountains and moors today is the crowberry, which avoids calcareous soils, and will grow well over limestone only when sufficient acid humus has accumulated.

This interpretation of the strikingly rich floras of places such as Teesdale has supplanted earlier speculations that these peculiar assemblages of plants are survivors of a flora which existed before the last, or even all, the ice ages. Such a theory of " per-glacial survival " which, so far as British vegetation is concerned, was copied from theories developed in Scandinavia and North America, may still, however, be necessary in some form to explain the remarkable distribution of certain Irish and western Scottish plants; but there seems to be no reason to think it applies to the vast majority of disputed cases. There is no doubt that, as our knowledge grows of the actual plants growing in this country in the Late Glacial period, we shall see more clearly the history of many of these remarkable distribution patterns of mountain plants today. These questions are treated in more detail in the book on *Mountain Flowers* in this series (Raven and Walters, 1956).

John Markham

Plate 15. Alpine Lady's Mantle, *Alchemilla alpina.* Cairngorms, September

a   *John Barlee*

b   *John Barlee*

Plate 16a. Moss Campion, *Silene acaulis.* Snowdon, June
    b. Starry Saxifrage, *Saxifraga stellaris.* Snowdon, June

CHAPTER 9

BOGS, FENS AND MARSHES
(M. W.)

To everyone living in the north or west of Britain the word *bog* is familiar as meaning treacherous ground carpeted with bog-moss (*Sphagnum*) and bearing a very special assortment of plants, including the cotton-grasses (*Eriophorum* spp.) and several members of the heath family (*Ericaceae*). Beneath its unstable, often quaking surface the bog has a variable depth of peat, and static water often stands at its surface. Peat is the largely undecomposed remains of plants which accumulate under conditions of water-logging, when the ordinary processes of decomposition which take place in well-drained soils are arrested. The typical bog peat is composed of layers of variable texture largely derived from the bog-mosses which have been the dominant plants growing there; and under favourable climatic conditions (such as the cool rainy climate of Ireland) peat-bogs may cover enormous areas and continue their active growth for hundreds or even thousands of years.

In the east of England, particularly in East Anglia and Lincolnshire, there were formerly extensive areas of peat deposits covered with a type of vegetation called *fen*, which differs from bog in that its characteristic mosses and ericaceous plants have played no significant part in the formation of the peat, which is accordingly of a different structure. This difference in vegetation is due to the effect of alkaline drainage water, for fens are formed in low-lying basins (which may have been originally lakes) in areas where the drainage water is coming from surrounding uplands with usually calcareous rocks and soils. Under these conditions the bog-mosses cannot grow, and an entirely different set of plants flourishes (see p. 140).

A third type of community developed on a water-logged soil is the *marsh,* in which the soil is a mineral one and development of peat is prevented, usually because of the existence of free drainage and erosion. Marshes thus develop naturally on slopes where streams or springs arise, or on the flood plains of rivers. The sea-shore *salt marsh* is a very specialised type of community which is discussed later (see p. 173).

BOGS

Peat-bogs are particularly characteristic of regions with a cool, rainy climate, and also form most readily in areas where the rocks are not calcareous. Since much of the north and west of Britain is composed of ancient, siliceous hills and mountains with a high rainfall, it is not surprising that peat deposits are widespread there; whilst in Ireland the extreme oceanic climate has caused the formation of enormous areas of bog in spite of the abundance of under-lying calcareous rock.

Three kinds of bogs can be distinguished for convenience—although, in an area of extensive peat development, it is impossible to separate neatly these different kinds. The three types are known as *raised, valley* and *blanket* bogs.

The *raised bog,* in its natural state, develops as an extensive peat area on more or less level ground; its growing surface is very slightly convex and is composed of a mosaic of very characteristic pools and hummocks, which may be only a few feet across. The bog grows by a process of peat accumulation in the pools, until the open water is replaced by bog-mosses; later, when the moss stands above the water level, other plants such as heather (*Calluna*), able to grow in the drier conditions, colonise the bog-moss. In this way a new hummock forms, and a new pool may originate below it on what was an old hummock, and the cycle is completed. This so-called " regeneration complex " can actually be seen, in raised bog peat which has been formed rapidly during an active growing period, in the form of lens-shaped areas visible on a vertical peat section; the main body of the lens is composed of the bog-mosses, whilst the upper boundary represents the hummock stage and contains remains of heather and other plants.

The peat, in fact, is able to provide a record of the growth of the bog—a record which, in the hands of trained investigators, can yield detailed information about the successive changes in the vegetation of the bog and its surroundings, and indeed about the origin of the bog itself. For this purpose a thin vertical section is taken through the peat by using a specially designed borer; the successive layers are then studied and their contents identified as far as possible. In this way it is possible to show that many raised bogs have, at their lowest layers, lake mud deposits and, immediately above these, fen peat formed largely from the common reed (*Phragmites*); then the bog-moss layers begin. Such a raised bog has obviously begun as a reed-swamp and fen (see p. 141) on the site of a lake. Not only the remains of stems, leaves, fruits and seeds can be identified in peat, but also, after suitable treatment, the microscopic dust-like pollen can be extracted, concentrated and identified. The recent development of this process of " pollen analysis " has given us a most remarkable insight into the history of the changes in vegetation during several thousands of years; a fuller account of this fascinating study and its results is given in Chap. 4.

The utilisation of peat for fuel, and, more recently, for cultivation, has, of course, resulted in the drainage and modification, or even destruction, of most of the large raised bogs formerly existing in Britain; and, even in Ireland, where raised bogs cover so much of the central plain, peat cutting and drainage leave their mark. Thus the large areas of Chat Moss and Carrington Moss west of Manchester have been drained and largely converted into arable land and, as in the Fens, only the names remain. " Moss," by the way, in the sense of " bog," which occurs frequently in the north of England, is a Scandinavian word which still exists in those languages to-day.

The *valley bog*, as its name suggests, owes its existence to water-logging and peat formation on the floor of a valley, particularly in regions where the drainage water from the surrounding hills is not calcareous. From a valley bog, under suitable conditions, a raised bog may in time develop. In the New Forest, where the parent soils are acid and sandy, there are well-developed valley bogs, some of which are famous as the localities for rare species of plant.

The third type of acid peat community is the *blanket bog*, char-acteristic of much of upland Britain (see also Chap. 8.) Here the peat forms a more or less continuous layer over the tops and gentle slopes of the hills, wherever the rainfall is high and distributed through-out the year, as on most northern hills. The surface vegetation of blanket bog is, however, usually poor in bog-mosses, and may be commonly composed of cotton-grass (see p. 82); such a vegetation cover may well be called moor rather than bog (see Chap. 6). Many of these blanket bog peat coverings show extreme stages of erosion. In the Pennines, for example, this erosion may have proceeded so far that over a considerable area no living vegetation can be seen, only bare peat and eroded drainage channels. Maps of streams on the plateau of Kinderscout differ by as much as half a mile in length between 1840 and 1940. Although this erosion may be in part due to climatic change, recent evidence (Conway, 1954) suggests that burning and grazing have played a very important rôle in the degeneration of the blanket bog. The end-point of this process can be seen on the summit plateaux of mountains such as Ingleborough (Yorks) or Bleaklow (Derbyshire), where the peat has now almost disappeared from large areas, leaving a bare stony or gravelly surface carrying a sparse and scattered vegetation of bilberry, crowberry and grasses. A study of the peat of typical Pennine blanket bog has shown that peat formation began in the Atlantic period about 8,000 years ago, and that there was in many places especially rapid peat accumulation in the more recent Sub-Atlantic period (see p. 54). Some of these deposits which are now eroding so rapidly may be up to 20 feet in thickness; immediately below the peat blanket there can be found in some places the evidence, in the form of tree-trunks and roots, of a former forest. Such forests, we think, date from a somewhat dry Sub-Boreal period, which was succeeded by a cool, wet Sub-Atlantic period at the time of the main Iron Age settlements in Britain, and in this climate, so favourable to peat formation, a mass of blanket bog peat was laid down.

In Ireland peat formation proceeds so rapidly and easily that blanket bog covers hill slopes of up to about 15°. Such bogs may be unstable, and if they reach a critical thickness may slip, causing the " bog-bursts " which occasionally take place on Irish hills, with disastrous results to farmland and cottages. Even over limestone, in Ireland, peat may accumulate, and eventually build up a blanket bog community in which no trace of the influence of the underlying rock remains; although in certain Irish limestone regions, notably the " Burren " of County Clare, this extensive peat formation has not taken place, and a rich limestone flora, rooted in a shallow soil, is developed (see p. 127).

The flowering plants which occur on these three types of bogs are

Plate XIII. Water Germander, *Teucrium scordium.* Cambridgeshire, August

M. C. F. Proctor

Eric Hoskin

Plate XIV. Bog Bean, *Menyanthes trifoliata*. Hickling, Norfolk, June

rather few, but contain some of the most curious and interesting members of the flora. The sundews (*Drosera* spp.) are striking and unmistakable plants; their neat rosettes of reddish leaves, studded with glandular hairs glistening in the sun, attract small insects which are trapped, enfolded by the hairs, killed and digested. Charles Darwin, whose gift of patient and accurate observation of nature enabled him to put together a vast body of evidence to support his theory of evolution of plants and animals through natural selection, wrote a book on insectivorous plants (1875), in which he recorded experiments with the sundews and many other species. There are three species of sundew in Britain, of which the round-leaved (*D. rotundifolia*) is the commonest, occurring throughout the country in suitable boggy ground, and usually actually growing in tussocks of bog-moss. The two other kinds, whose leaves are not circular in outline, are more restricted in their occurrence. The smaller of the two, *D. intermedia*, is not uncommon in English bogs and wet heaths, but is usually to be found growing on bare peat, often by little drainage channels, rather than in the bog-moss tussocks. In many bogs, such as the valley bogs of Norfolk and the New Forest, both species grow near together but preserve this difference in habitat preference. The third species, *D. anglica*, (Pl. XI, p. 110) which is usually larger than either of the others, has a rosette of very long spathulate leaves which are held more or less erect. It is rather unfortunately named, for although it *does*, as its name suggests, occur in England, it is actually much commoner in Scotland and Ireland. Where it grows with the round-leaved sundew, hybrids are sometimes to be found. The flowers of the sundews are nearly always cleistogamous; that is, they do not open fully and are self-pollinated in the bud (see p. 34).

There are two other types of insectivorous plant to be found in the British Isles, namely the butterworts (*Pinguicula* spp.) and the bladderworts (*Utricularia* spp.). The common butterwort (*P. vulgaris*), like the sundew, has a rosette of leaves; but these are large, fleshy and yellow-green, and catch small insects with a sticky secretion on the upper surface. The flower of the butterwort has a beautiful, large-spurred purple corolla and is borne singly on a long slender stalk in June and July. The plant persists in the form of rootless buds in winter. Neither the common butterwort nor the common bladderwort (*U. vulgaris*, Pl. 17a, p. 146) is typical of acid bogs, in spite of many and persistent statements to the contrary to be found in books.

The butterwort occurs throughout England (though it is rare in the south) in marshes and fens and wet peaty heaths where the water is not too acid; it does *not* normally grow in bog-moss, and where it occurs together with sundew, it will usually be found growing in the wet seepage channels from which the bog-moss is excluded. Even in the north and west of the British Isles, where both sundew and butterwort are common, this difference in preference tends to remain. In the East Anglian fens, where both sundew and butterwort are nowadays very rare or extinct, their rarity is not due to the fact that their requirements are the same, but more simply to the destruction through drainage of all suitable ground. We know from old herbaria and other records that sundew once grew abundantly at a number of places on the fringe of the East Anglian fenland, particularly around the large meres north of Huntingdon in the area where the Wood Walton Fen Nature Reserve is now situated, and it still grows on wet acid heaths and small bogs at or near the fenland fringe in west Suffolk; but we also know that in the Wood Walton area there flourished, probably until the nineteenth century drainage operations, one or more large acid bogs marginal to the fenland, in which the bog-mosses, sundews and all the typical bog plants flourished. Drainage of the meres and surrounding land here has converted the remnants of the bog into an acid heath, more or less dry in summer, on which are only fragmentary relics of the former vegetation, and the sundews have long disappeared. With the butterwort it is, however, a different matter; its present-day rarity in the Fens is due, firstly, to continually extended drainage and cultivation of all the small fen remnants which persisted in the nineteenth century; and, secondly, to its inability to compete with taller vegetation; it grows well in alkaline fens and marshes if it is not " crowded out."

Three other species of butterwort are included in the British flora, though one, the alpine butterwort (*P. alpina*) has not been re-discovered for very many years and is thought by many botanists to be extinct. The other two butterworts are western plants. Pale butterwort (*P. lusitanica*, Pl. XII, p. 111) is not uncommon on acid, boggy moorland in Cornwall, Devon and Hampshire; from its pale rosettes arise delicate pale lilac flowers borne singly on slender stalks. The Irish butterwort (*P. grandiflora*) does not occur in Britain, and is a very good example of a so-called Lusitanian species, common in northern Spain and the western Pyrenees and occurring also in the west of

Ireland. It bears a large, handsome deep purple flower with a very long spur; where it grows in Ireland with the common butterwort. many intermediate plants, presumably of hybrid origin, can be found,

The bladderworts are submerged water plants growing in peaty pools and possess very remarkable " bladder-traps " which engulf small water-creatures and digest them. They flower in July and August, sending a slender stalk up above the water surface which bears several beautiful two-lipped and spurred yellow flowers. The common bladderwort (*U. vulgaris*) occurs in fen ditches, ponds, etc., scattered throughout England; the other species are local and rare in their occurrence. One of them, *U. intermedia*, which grows in shallow peaty pools or wet mud in fens and wet heaths, bears its bladders on special separate branches; it is very shy of flowering, and so is generally supposed to be even rarer than it really is.

Remarkable though they are, the insectivorous plants are never very important constituents of bog or fen vegetation, and we must now turn our attention to some of those grass-like plants which form the dominants in bogs. It is interesting that most of these belong to the sedge family, *Cyperaceae*. Most (but not all) sedges are easily distinguished from the true grasses (*Gramineae*) by having their leaves arranged in three planes, seen from above; or, if no proper leaves are developed, the stem is seen to be triangular in cross-section; the grasses tend to have leaves in two rows and a more or less round stem.

Grasses and sedges are dull plants to most amateur botanists, for with few exceptions their flowers are small and inconspicuous and adapted to wind-pollination. Because of their importance in agriculture the professional botanist, at least, usually has to learn to recognise the commoner grasses not only by their flowering heads, but also from their leaves and stems; but the sedges have no such economic importance, and so are relatively little known. This is a pity, because with a little practice, most of the commoner sedges can readily be distinguished from each other, and show many points of interest, both in their structure and their ecology. Most of our sedges belong to the large genus *Carex*, whose members vary in size from tiny plants only a few inches high to giants reaching six feet or more, and include some of the commonest and also some of the rarest plants in the British Isles. The fenman's " sedge," however, which on Wicken Fen, for example, has been regularly harvested and used for thatching for several hundred years, is not a *Carex*, but belongs to a separate genus, *Cladium*. This *Cladium* was once a very abundant plant in the East Anglian fens, but drainage and cultivation have now reduced it to a few preserved fens and small, partially undrained areas. It is an evergreen plant whose long rigid leaves, armed with saw-teeth on the margin, grow for several years from the base; infrequent cutting will not seriously harm the plant, but if it is cut annually, the strength of the plant is seriously reduced, and (on Wicken) the grass *Molinia coerulea* tends to replace it. The old fenmen knew this; and the regular practice in cutting sedge-fields was to take a crop every fourth year, which is still done at Wicken.

Cladium is never found on acid bogs, which have a different set of characteristic members of the sedge family. The best known of these are the cotton-grasses (*Eriophorum* spp.), which we have already mentioned. Two kinds of cotton-grass are common over wide areas of bog, particularly in northern Britain; these are the single-headed tussock-forming *E. vaginatum*, (Pl. 10*b*, p. 83) and the several-headed *E. angustifolium*. Both are striking plants when in fruit, with their conspicuous white cottony heads. The cotton consists of very long white bristles situated beneath each small seed-like fruit; these bristles, and similar structures in related plants, are generally thought to represent the perianth of the tiny flowers.

The cotton-grasses well illustrate how plants similar in appearance may yet differ very markedly in their requirements, for there is a less common species, *E. latifolium*, which resembles *E. angustifolium* closely, but which occurs in marshes where there is alkaline drainage water, and never on really acid peat. The other British cotton-grass, *E. gracile*, is a very rare plant of a few bogs in Hampshire and other southern and midland counties.

One of the difficulties which confronts the student who tries to learn some of the British sedges is that it is not easy to decide from Floras what are the *commonest* species which he is most likely to meet in particular types of vegetation; this means that he may spend a great deal of time considering whether a common plant he has found is one of the many local or rare species included in the Flora. The following short list of the commonest *Cyperaceae* may be helpful, arranged according to type of habitat:

1   Meadows, roadside-grassland: *Carex hirta*
2   Fens, alkaline marshes: *C. flacca, C. disticha, C. elata, C. panicea*

3. Calcareous grassland: *C. flacca, C. caryophyllea, C. panicea*
4. Reed-swamp fringing ponds, ditches, etc.: *Eleocharis palustris, Carex riparia, C. acutiformis, C. otrubae, Schoenoplectus lacustris*
5. Coastal: *Carex extensa* (salt marshes, etc.); *Carex arenaria* (sand-dunes)
6. Woodlands: *C. sylvatica*
7. Acid heath, moorland and bog (lowland and upland Britain): *Carex nigra, C. binervis, Trichophorum caespitosum, Eriophorum* spp.

All these plants are easily recognisable in fruit; when once they are known, one has a useful basis on which to build a greater familiarity with the group. Two of these commonest sedges, *C. flacca* and *C. nigra*, occur not only in several different kinds of habitat but also can vary a good deal in size, shape and colour of inflorescence; this is important to remember when trying to identify sedges. The others do not vary much, either in appearance or in their choice of habitat.

Let us turn back now to the plants growing in bogs, and consider some of the more attractive species. The heath family, which we have earlier mentioned (see p. 86), supplies several, notably the delicately beautiful cranberry (*Oxycoccus*), whose slender trailing stems may be found in bog-moss, bearing small pink flowers with reflexed petals (like cyclamens) and later, the familiar fruit. Actually the usual cranberries of commerce are larger than our wild ones, and come from a different, North American, species (*O. macrocarpus*); but they are closely similar. In parts of the north of England the fruits of the cowberry (*Vaccinium vitis-idaea*), which is abundant on many Pennine moors, are sold as "cranberries"; they are good to eat stewed or as jam, but their flavour is not that of the true cranberry. Cranberries are not easy to pick, as anyone who has tried will know, for they ripen amongst the bog-moss, with very slender stalks bending under the weight of the berry, and must be picked out at ground level in what is often a soft and squelchy bog—but they are very rewarding when converted into cranberry jelly. A curious fact about the cranberry is that a small population often contains a mixture of differently shaped and coloured berries, each type being borne, of course, on a different plant. Some berries are round, others pear-shaped; and the colour when ripe may be a rich red, or a curious speckled brown-red. All four possible combinations of shape and colour may occur in one bog. Over a very small area, however, where all the berries are borne on what were branches of one original plant, the type of berry is uniform.

Another interesting ericaceous bog plant is *Andromeda polifolia*, which, though formerly abundant on many raised bogs in the north of England, is now greatly reduced through drainage. This is a stouter, woodier plant than the cranberry, with larger leaves and flowers, and a dry capsular fruit. In Sweden and Finland, where there are very many bogs, another " heath " is very common, *Ledum palustre*, an aromatic shrub much used in herbal medicine. The occurrence of this plant in Britain is mysterious. It has generally been thought to be introduced, but in at least one of its Scottish localities it is growing on a raised bog in a very natural way, and the evidence for its introduction there is not strong. *Ledum palustre* has been grown in English gardens for a long time as " Labrador tea."

One does not associate the lily family (*Liliaceae*) with bogs—or even with wet places in general; but there is one British member which is at home on acid bogs and often locally abundant there. This is the bog asphodel (*Narthecium ossifragum*). The beautiful orange-yellow flowers appear in July, and contrast vividly with the greens of bog-moss in which the plant so often grows. Unlike most *Liliaceae*, which are bulbous, the bog asphodel has a creeping rootstock, and spreads easily through the loose bog-moss to form large patches.

Finally, we might mention *Scheuchzeria palustris*, a plant which frequents only the wetter parts of acid bogs, and which, because of its rarity and the rather dull rush-like appearance of its flowers, has no English name. Formerly this plant grew in several parts of England, but any drainage is likely to eliminate it, and it is now almost confined in Britain to the Moor of Rannoch, a fantastically desolate stretch of bog-covered country through which runs the main railway line to Fort William. Here, in an uninhabited waste of bog, pools and moor-land, *Scheuchzeria* is still very much at home. *Scheuchzeria* is a curious plant, with no very obvious close relatives; it is often included with the arrowgrasses (*Triglochin* spp.), or put in a special family of its own.

Fens

Fens, as we have seen, resemble bogs in that the vegetation cover has formed, and may still be forming, peat layers, but the soil water is not acid, and the characteristic species of bogs (the bog-mosses, the *Ericaceae*, and certain sedges) are absent. As one would expect, fens are characteristic of the lowland parts of Britain where the softer, generally more basic rocks occur, and are (or were before drainage) best developed in East Anglia. We know that at the time of Hereward the Wake and the Norman Conquest, the Isle of Ely really *was* an island, access to which (from adjacent raised ground) was gained by artificial causeways across the impenetrable fenland. In winter such fenland must have been quite impassable, particularly when high tides banked up the water of the sluggish rivers draining into the Fenland basin from the inland chalk of Cambridgeshire, Huntingdon-shire and other adjoining counties. Exactly how effectively the Romans drained the Fens we do not know; but there is evidence that their work was very considerable, and that after their departure it fell into disuse for a thousand years or more. The seventeenth century saw the beginning of the modern drainage system, which, since that time, has gradually converted almost the whole area into farm land, some of it the richest in the country. A few remnants have fortunately been spared as nature reserves, and in at least one of these (Wicken Fen) we have good reason to believe the vegetation now is not radically different from what it was all over the area before the seventeenth-century drainage schemes began.

To understand the fenland vegetation we must consider what happens to any stretch of shallow water left undisturbed. Water plants (see Chap. 10) will establish themselves in it, and round the margin will spread a reed-swamp in which the likeliest plant is the common reed (*Phragmites communis*). As the reed-swamp extends, the annual remains of the stems of the plants will die down and accumulate under the water, to form, as time goes on, a layer of peat. Gradually this peat will build up to the water surface, and new plants will come in which are able to compete with the reed-swamp and gradually to beat them in the somewhat drier conditions. If the pond is very shallow, then, it will, over a period of years, be converted first into reed-swamp and later into some type of fen community. At Wicken this would be dominated by the sedge *Cladium* (see p. 138). But the process should not naturally stop there; plant remains would still accumulate (though peat itself would not form once the water-level was below the soil surface), and trees or shrubs would invade the fen (see p. 76). The final product of this natural succession or *hydrosere* (see p. 39) would be some sort of woodland, under the shade of which almost all the fen species would have been killed out. A fen, therefore, is not such a permanent community as a raised or blanket bog may be; and if fen persists in a region for a very long time, we must look to complicating factors to provide an explanation. The most obvious one is that of human interference; and it is to regular cutting of the sedge and " litter " that Wicken Fen owes its preservation. In that sense, therefore, Wicken does not represent a natural vegetation; if we allowed the vegetation to change naturally there we should be left with a woodland, not a fen. Why, then, have large areas of fenland survived for so long—as we know they have—in different periods in the past? The answer to this appears to be that since the ice age, not only the climate but also the relative levels of land and sea have changed very considerably, and that at certain times a wet climate has coincided with a rise in sea-level relative to the land adjoining the fenland basin. Such a period seems to have set in towards the end of the Roman occupation of Britain. This meant that, particularly after the collapse of Romano-British civilisation, the Fenland basin was inundated and great shallow lakes were formed. Some of these " meres," notably Whittlesey Mere, were not drained until the middle of the nineteenth century, and as on parts of the Norfolk Broads to-day, around their margins were enormous areas of reed-swamp and fen. Of course, the opposite process has also occurred in the history of the Fens; as any Fen farmer knows, " bog-oaks," many of enormous size, are frequently found buried in the peat, and some of them date from a very early period when the fen surface dried out considerably and a great forest spread over it. But in later periods the water came again and the remains of the forest were buried in peat. In the layers of peat, as we have seen (Chap. 4), is preserved a revealing record of the vegetational history of the Fens.

Fen does not necessarily develop into woodland. If the climate is wet enough, as in the central plain of Ireland, bog will readily take over from the fen communities which form in shallow lakes on this predominantly limestone region (see p. 134). Indeed this may happen also in the drier climate of East Anglia in particular situations. In Norfolk to-day there are a good many fens in which tussocks of bog-moss occur locally.

We have already mentioned the Fenman's " sedge " when talking of sedges in general. Let us now consider some other characteristic

M. C. F. Proctor

Plate XV. Sweet Flag, *Acorus calamus*. Cambridgeshire, July

Plate XVI. Lesser Water-plantain, *Baldellia ranunculoides*. New Forest, June

fenland plants, many of which, though formerly common, are now rare or local. At Wicken one of the most famous is the milk parsley (*Peucedanum palustre*—the " carrot " of the Fenman), a tall white-flowered umbellifer on which the larvae of the swallow-tail butterfly feed. This occurs quite commonly at Wicken, particularly in areas of the mixed fen which comes into being after the clearance of young " carr " of alder-buckthorn (*Frangula alnus*, see p. 76). It is a noble plant, standing several feet high and bearing large numbers of white flowers in late July and August. It is a very curious, and as yet quite unexplained, fact that at Chippenham Fen, the other Cambridgeshire reserve, only a few miles from Wicken, the locally common umbellifer in the sedge and mixed fen vegetation is superficially similar to the milk parsley, but is on closer examination seen to be entirely different. This Chippenham plant is, indeed, a rare plant in Britain, and is called *Selinum carvifolia*.

The history of its discovery in this country is very interesting and points several morals. It was first found in a marshy meadow at Broughton in Lincolnshire in 1880 by the Rev. W. Fowler, and then collected the following year at Chippenham by W. R. Cross, an amateur botanist at Ely. The Cambridge botanists of that time, refusing to believe that they had overlooked such a large and distinct plant in the county, considered that it must have been recently intro-duced—a view for which there has never been the least shred of evidence! In 1910 it was discovered in Nottinghamshire; but in recent years, possibly through drainage, it seems to have gone from both this and the original Lincolnshire locality, and so far as is known is now confined to Cambridgeshire. Recently it has been found in some quantity in another grazed fenny meadow in Cambridgeshire, several miles from Chippenham. The reasons why it was so overlooked in Britain would seem to be firstly its superficial resemblance to *Peucedanum palustre* in flower, secondly its very late flowering season, and thirdly, in the absence of flowering stems, the difficulty in dis-tinguishing the leaves from those of several umbellifers with finely dissected leaflets, such as *Peucedanum* itself, the wild carrot and *Silaum silaus*. Indeed, its recent discovery in a quite unexpected Cambridge-shire locality encourages the belief that *Selinum* may well be elsewhere undetected—particularly, for example, in the fens of the Norfolk Broads.

Another typical fen plant, well known at Wicken, is the marsh pea

(*Lathyrus palustris*). This is a beautiful species climbing with twining tendrils on the tips of its leaves amongst the sedges and grasses of the fen. It flowers freely in July and sets abundant seed in pods. The fen violet (*Viola stagnina*, see p. 93), abundant at Wicken in the 1860's, has, however, completely disappeared; though fortunately it still flourishes at Wood Walton Fen, and in a few scattered fens in several eastern counties—and also in the peculiar " turlough " vegetation of the limestone of County Clare in the west of Ireland.

One of the most famous Fenland rarities is the fen orchid (*Liparis loeselii*). This was once so abundant on Burwell Fen, that in 1835 the child secretary of the Swaffham Prior Naturalist's Society could write: " We had very good sport both in plants and insects. *Ophrys [Liparis] Loeselii* was found in great plenty. Between four hundred and five hundred specimens were brought home. It was growing in the grass and moss among the pits where they cut turf. There were two bulbs to each plant, and the bulbs were scarcely in the ground at all, so that we picked them out easily with our fingers."

It is now a very rare plant in the whole of England. Although the unprincipled over-collecting of the nineteenth-century naturalists must have contributed to its decline, it is unlikely that this has been the most important reason. The fen orchid well illustrates the fact—known to most field botanists—that the richest fen flora is to be found in places where there is a variety of disturbances, such as is caused by peat-digging, rough grazing by cattle, or occasional burning and cutting by man. In such places the normal hydrosere succession is suspended or deflected, and we get a number of fragments of all sorts of stages of development, from open peaty water in old peat cuttings, to the " mixed fen " which is more or less stabilised by burning and rough grazing. Now, very many fen plants of small stature, such as the fen orchid, are excluded by competition from the tall-growing reed-swamp and fen, and have no place in the ordinary succession. Their chance comes when some accident—usually attributable to man—creates an open more or less bare patch of peaty ground in which they can temporarily flourish, till the tall dominants close in again. The fen orchid flourishes in damp, but not shady, hollows in rather " open " fen, in which the " sedge " and other tall plants are somewhat patchy; and of course, cutting, rough grazing or trampling can keep such hollows in existence. Thus drainage alone is not responsible for the disappearance of *Liparis*; but rather drainage and a cessation of disturbances. To many people it might seem easy to protect a particular rare plant by putting a fence round it; but all too often this would merely hasten its demise, for most of our rare species are rare precisely because they do *not* manage to fit in to the normal natural closed vegetation, but eke out a precarious existence in often very temporary habitats. And this is nowhere better seen than in the fenland flora. The fen orchid also occurs, and is locally abundant, in a few dune slacks on the coast of south Wales. This type of distribution—coastally and on inland fens—is shown by a good many plants, and is bound up with the question we have been discussing. (Other good examples are the brookweed, *Samolus valerandi*, and the creeping willow, *Salix repens*, both of which are widely distributed around the coasts in marshes and dune slacks, and also occur at Wicken.) These sea-coast habitats (sand-dune slacks, brackish marshes, even sea-cliffs) provide places where tall dominants are excluded by the particular conditions —e.g. exposure (especially to salt), instability of soil, etc.—and in them we find this type of plant, which leads a much more precarious existence in inland marshes and fens. Yet we know that there was a time in the history of the flora when such plants must have been abundant; this was in the Late Glacial period.

The so-called " marsh orchids " are robust plants with thick spikes and unspotted leaves. The most widespread species is *Orchis strictifolia* (Pl. 17*b*, p. 146) which occurs in marshes, wet meadows, and fens throughout the British Isles. Its flowers are usually, in the typical plant, pinkish or flesh-coloured.

As we saw earlier, there are rather few species which will tolerate the dense shade which is developed as the fen is colonised by bushes and trees. One of these is particularly interesting, however, as it is a fern. The marsh fern (*Thelypteris palustris*) must have been at one time very abundant indeed in the East Anglian fens; it grows quite well in thick sedge, and very luxuriantly under young bushes in which the sedge is dying, and spreads very rapidly by means of a long rhizome which runs along in the top few inches of peat. No other fern appears to be able to thrive in fen peat under trees; and, although several occur very commonly on woodland, it is usually on neutral or acid soils.

MARSHES

The distinction between marshes on the one hand and bogs and fens on the other is, as we have said, conveniently based on the absence of any considerable peat soil in a marsh, which is often developed on a slope or in some other position where the accumulation of peat is prevented (e.g. by river scouring or erosion). Fragmentary marsh communities occur, of course, throughout the country, but in lowland Britain they are nearly always highly modified by grazing or other factors. The commonest lowland marsh (which may grade into fen and bog communities) is that in which rushes (*Juncus* spp.) are dominant, often on a gentle slope or in meadowland in the alluvial plain of a river. Common plants found in this type of grazed and disturbed marsh include the creeping buttercup (*Ranunculus repens*), the flea-bane (*Pulicaria dysenterica*), and the coarse, tussocky hair-grass *Deschampsia caespitosa*.

In upland Britain, less disturbed marshes are more common; but, as we have already seen, the whole of Britain, including the mountains, is pretty effectively grazed, so that the grazing factor is usually present even in marshy patches on high mountains. In areas where the underlying rock is acid, accumulation of peat is very easy, and a gentle slope may carry a blanket bog. Steeper slopes where erosion is operating may show a marsh with species of rush (*Juncus*), lesser spearwort (*Ranunculus flammula*), marsh violet (*Viola palustris*), and other plants. In regions of limestone or other basic rocks, on the other hand, marshy slopes may carry a very rich, luxuriant vegetation. Thus on the Yorkshire and Derbyshire limestone, plants such as the globe-flower (*Trollius europaeus*), and the great burnet (*Sanguisorba officinalis*) occur on marshy slopes; and in certain areas, where the marsh vegetation is less luxuriant and more open (as in parts of the Craven limestone in the Malham area), there may be a varied assortment of smaller species, many of which also occur in " open " ground in fens, e.g. grass of parnassus (*Parnassia palustris*), and butterwort; but others, e.g. the bird's-eye primrose (*Primula farinosa*), are of a more northern type. Even in such places, acid peat can easily begin to form, and small tussocks of bog-moss, with sundew and other plants, raise themselves above the general level of the marsh and therefore out of the main influence of the alkaline drainage water. Once a thin layer of peat is established, it will accumulate quickly, and bog plants

a   *Eric Hosking*

b   *S. C. Porter*

Plate 17a. Bladderwort, *Utricularia vulgaris*, flowering in peaty pool. Wicken Fen, Cambridgeshire, August

b. A Marsh Orchid, *Orchis strictifolia*, and Marsh Horsetail, *Equisetum palustre*. Suffolk, June

a

b

Plate 18a. White Water-lily, *Nymphaea alba*. Wicken Fen, Cambridgeshire, August
b. Arrowhead, *Sagittaria sagittifolia*. Kew, August

will thrive; but winter erosion or some other factor may break up the peat cover again and wash it away. Such areas of calcareous marsh are often mosaics of plants requiring very different soil conditions.

Bog and fen plants have suffered more than most other members of the flora from the recent spread of man's activities; for even partial drainage will cause a rapid change in the surface vegetation and will eliminate many rare and interesting plants. The case for fen nature reserves is clearer than most, and we may hope that at least in a few remaining places these beautiful and interesting plants—and their associated animals—may be permitted and encouraged to survive.

CHAPTER 10

RIVERS, LAKES AND PONDS

(M. W.)

THE COUNTRYSIDE of Britain offers varied delights to a field botanist; he can seek the satisfaction of a strenuous day on mountain and moor, then turn for contrast to easier, perhaps more idle, pleasures by the ponds and gently-flowing streams of the lowlands. Some of our finest wild flowers are to be found there—the white and yellow water-lilies carpeting the surface of the still water, and the purple loosestrife, yellow iris or blue water forget-me-not fringing the banks where the shade of the willows and alders is not too deep.

Life in fresh water presents its peculiar problems to those flowering plants which have adopted it. The simplest forms of green plant-life that we know—the unicellular Algae—are, with few exceptions, to be found in water, both salt and fresh; and there is very good reason to believe that plant life began in the sea and later conquered the land. Thus all the more elaborate " higher plants," from mosses and ferns to the flowering plants, are really land plants, adapted to growing with their roots in the earth and their green parts in the air. However, just as in the insects and mammals, we find that certain types, such as the water-beetles and the whales, have " gone back " to the water, so we have a group of higher plants which are secondarily aquatic. We know that these flowering plants have had land ancestors, in some cases because they are closely related to whole groups of land plants, but in general because their method of sexual reproduction involves the formation of flowers and the mechanism of pollination which (with very rare exceptions) can and does only work if the flower is produced *in the air*. The Algae, in contrast, show types of sexual reproduction which are dependent upon *free water* surrounding the plant.

The flowering plants of rivers and lakes show several interesting features connected with obvious differences between a water and a land environment. In the first place, they tend to have soft stems with large air spaces and very little strengthening tissue—there are no fully aquatic trees or shrubs. And secondly, most of them flower and set seed far less regularly and abundantly than most land plants, depending very largely for their spread on vegetative methods. This is, one assumes, connected with the difficulty which the water plant may have in producing regularly flowering stems which succeed in reaching the air. Some, of course, which are adapted for life in still waters (such as water-lilies), succeed in producing their large showy flowers every summer on the water surface, and set seed; but very many water plants flower irregularly and fruit even more irregularly. For the same reason, presumably, annual water plants are rather rare; and most of them are not strictly aquatic, but rather plants of open muddy habitats, e.g. water-blinks (*Montia*) and mud-wort (*Limosella*). We can in fact distinguish for convenience three kinds of water plants; these are the free-floating (e.g. duckweed, *Lemna*), the submerged (usually at least loosely rooted, e.g. Canadian waterweed, *Elodea*) and the rooted type with floating or aerial leaves (e.g. water-lily, *Nymphaea*, Pl. 18*a*, p. 147) and bog-bean, *Menyanthes trifoliata*, Pl. XIV, p. 135). This last type cannot, of course, be sharply distinguished from the plants of the reed-swamp proper (e.g. common reed, *Phragmites*) which we have mentioned in Chapter 9.

Let us consider now the sort of habitats which are available for these water plants in Britain. We have seen earlier (p. 113) how, broadly, on both climate and geology, Britain can be divided into an upland north and west with predominantly acid soils, and a lowland south and east with mostly neutral or basic soils. This, of course, has a marked effect on the water plants. Lakes and ponds in north and west Britain tend to have " soft " or acid water, and the plants found in them are, therefore, generally calcifuge and not those commonly found in the lowland south and east. Sterile stony and gravelly shores are also characteristic of upland lakes in northern Britain; whereas in the south (e.g. the Broads) the lake shore (except in the case of artificial reservoirs or where recently artificially cleared) is a marginal reed-swamp shading into fen, and small shore species would have no chance to compete with the tall reeds. Again, streams and rivers in upland Britain provide habitats of swiftly running water very different from those of the meandering rivers of the lowland. As we would expect, free-floating plants (e.g. duckweed), and plants with large floating leaves such as the water-lilies, are restricted to still or very slowly-moving waters, whilst in the flowing waters we find plants with long thin and submerged stems and leaves which sway in the currents. It is interesting that the same plant when growing in still or moving water may look quite different; thus the bulrush (*Schoeno-plectus lacustris*), typically a tall reed-swamp plant, occurs commonly in not too swiftly-flowing rivers in a submerged form with long soft strap-shaped leaves, and never flowers. The same is true of the bur-reed, *Sparganium ramosum*.

This plasticity, or the ability to adopt quite different growth-forms in different environments, seems, indeed, to be characteristic of most aquatic plants. One of the most interesting groups which might be studied from this point of view is that of the white-flowered water crowfoots or "batrachian Ranunculi" (Pl. XVII) which occur in all types of fresh water from stagnant muddy or peaty ponds and ditches to quite swift-flowing streams. The identification of these water crowfoots is a matter of some difficulty, and they constitute what systematists call a "critical" group, in which the limits of species and subspecific units are not at all clearly defined, and which, therefore, different authorities have tended to treat in different ways. Broadly it is possible to distinguish three kinds of crowfoots; those which normally grow on damp mud rather than submerged in water and which produce only the lobed "floating" type of leaf (e.g. *Ranunculus hederaceus*, the "ivy-leaved crowfoot" common in southern England); those which normally grow in fairly shallow and still water and may produce both finely-divided submerged leaves, and lobed floating leaves (e.g. *R. aquatilis* and its subspecies), and finally those which grow in deep or swiftly-running water which rarely or never produce anything but submerged finely-divided leaves (e.g. *R. circinatus* of deep fen ditches, and *R. fluitans* of flowing rivers). Clearly the ability to produce both types of leaves is, on the whole, possessed only by those species which normally occur in situations where the plant is submerged but reaches the water surface to produce the floating type of leaf. Such plants, if grown on wet mud, will tend to produce very few submerged leaves; conversely, if grown in deep water, they may produce no floating leaves. Which type of leaf is produced must, then, in these cases, be determined at an early stage of development

Plate XVII. Mud Crowfoot, *Ranunculus lutarius*. New Forest, June

of each leaf by some conditions of the environment such as, for example, oxygen content or light. In the ivy-leaved crowfoot, however, the ability to produce submerged leaves is not there, for if a plant is grown in water, it will still produce only the floating type; conversely, a plant of *R. circinatus* cannot be induced to produce floating leaves by growing it on mud or in very shallow water. We see, then, in the water crowfoots different degrees of genetical control over the form of the plant; in some the control is so rigid that no matter what the environment only one type of leaf is produced; whilst in others the genetical control is laxer, and the environment determines whether both leaf-types are produced, or only one of them. Recent experimental work by Cook and others is conveniently summarised in Briggs and Walters (1969).

Many fresh-water habitats differ in an important respect from, say, woodland and moorland; they may be very temporary indeed. We are all familiar with the fact that a shallow ditch, newly cleaned, soon becomes grown over with reeds and other plants. This is, of course, a natural process—the hydrosere—which we have discussed in Chapter 9. The fact that this will tend to happen in any stretch of open water means, then, that there is an impermanence about the habitats of water plants in general, if one compares these with natural woodland habitats, for example. All fresh-water plants, in fact, can be looked upon as inhabiting particular stages in successions from open water to dry land. Naturally the *rate* at which succession will go on in a shallow pond is very much greater than in a deep flowing river; and, of course, in swiftly-flowing water erosion may permanently prevent the establishment of normal hydrosere stages. In deep lakes, again, particularly where the shore is stony and the water base poor (as, for example, in many Scottish lochs), the succession may be virtually non-existent. It is, however, useful to consider the process as the normal one which in certain circumstances is retarded or arrested.

The influence of man in altering, creating, and destroying fresh-water habitats in this country is very great indeed; and it is roughly true to say that in most of lowland England almost all the still water habitats are artificial ponds, reservoirs, canals and ditches. Even

Plate XVIII. Lupins, *Lupinus nootkatensis,* naturalised on shingle by River Tay, Perthshire.   (*R. M. Adam*)

the Norfolk Broads, which until recent years have always been considered natural features, prove on detailed investigation to be largely if not wholly the result of a mediaeval peat-digging industry which flourished on a vast scale (see Ellis, 1965). Large meres famous for their water-fowl existed until the eighteenth and nineteenth centuries in the Fenland area also, but these were drained for agriculture (see p. 141). The rocky upland lakes and tarns of the north and west, however, and the Scottish sea-lochs, are hardly affected by man's handiwork; they owe their existence almost entirely to the action of ice during the ice age, and their vegetation, we may assume, has developed uninterruptedly since the ice last retreated. Such lakes provide in their mud sediments a valuable historical record similar to that provided by peat bogs; in particular cases we know that where a bog is now situated was formerly a glacial lake (see p. 133), and in some places (e.g. the famous Malham Tarn) we have at the present day both peat bog and open lake, the former developed in the shallow end of the lake basin. Such upland tarns on limestone areas provide a rich variety of habitats; not only is there an interesting flora of the tarn itself, but also the surrounding communities show a rich variety, from an open gravelly margin on the exposed side, to fen and bog vegetation on the sheltered side. In contrast, a small stony upland tarn in siliceous mountains (such as are found throughout the Lake District) may have a very poor flora and little or no fringing reed-swamp. Certain very interesting water plants (e.g. water lobelia) do, however, occur commonly in these stony upland tarns and will be mentioned later.

Some stages may not be on the main line of succession, but may survive precariously, usually because of some interference; a good example is the muddy trampled margin of a pond to which cows have access, on which a number of water plants may be commonly found (e.g. the small flote-grass, *Glyceria declinata*, the water-pepper, *Polygonum hydropiper*, and others). Such low-growing water or mud plants seem to be unable to compete with the tall reeds of the reed-swamp and thus occur only where the tall plants are kept out. Extreme cases of this are afforded by plants such as *Elatine*, tiny inconspicuous water plants, usually completely submerged in very shallow water at a pond or lake margin. The impermanence of a still fresh-water habitat means also that the first arrival at a newly dug pond or ditch, for example, may have a great advantage over later arrivals—in other words, chance dispersal of seeds (or other parts of the plant which may root) seems to play quite an important part in determining which plants are abundant in a particular pond. We have already mentioned that, while many water plants flower and seed irregularly, most have very efficient means of vegetative spread; and this is of the utmost importance in establishing the plant quickly in a newly available habitat. Professor Godwin, in an early study on the vegetation of several artificial ponds in the Trent valley (Godwin, 1923), showed that the considerable differences in their flora were best explained along these lines.

What are the agencies which would bring about the introduction of water plants to a new pond or ditch? Very few water plants have light wind-blown seeds; and much the likeliest method of spread would seem to be in the form of seeds and fragments of plants attached to water fowl. Guppy (1893—see Arber, 1920, p. 373), who studied this problem in some detail, obtained considerable numbers of seeds of water plants, many of which germinated, from the guts of wild ducks, and emphasised the importance of this means of dispersal ; and there is some evidence to support the view advanced by Darwin in the *Origin of Species* (pp. 383-8) that seeds of water plants may easily be conveyed in mud adhering to the feet of wading birds. This is a fascinating topic, and in spite of its interest and importance we still do not have much real evidence; and thus Mrs. Arber's plea (1920, p. 300) that local natural history societies could contribute valuable observations over a series of years might well be echoed to-day.

It is at first sight a rather remarkable fact that water plants are on the whole much more widely distributed over the earth's surface than are land plants; for we might expect that enormous barriers of salt water and dry land would prove insuperable. The fact remains that many aquatics and semi-aquatics occur in suitable habitats in all five continents (e.g. the duckweed *Lemna minor*) or range from the North Temperate zone into the Tropics (e.g. the reed *Phragmites*). If we assume, with most biologists, that species, or at any rate the vast majority, have had one centre of origin, we must also, it seems, assume that long-distance dispersal of seeds and fragments of water plants is quite efficient, so that over very wide areas a suitable habitat will tend to contain a particular kind of water plant. It looks, therefore, as if the ranges of most land plants (whose chances of dispersal within one continent, for example, one would on the whole rate higher than those of aquatics) are limited not so much by failure of dispersal, as by the absence of suitable habitats. There is indeed good reason to think that fresh-water habitats, even if much more scattered, may well be much more nearly uniform (in temperature range, etc.) over a wide area of the earth's surface than land habitats. In other words, plants *are* reasonably successful in dispersing themselves, and the general difference between the ranges of water and land plants merely reflects the greater uniformity of the fresh-water habitat.

To say that plants are reasonably successful, however, is not to forget that we have many striking examples of plants which, on being introduced by man from a distant country, either accidentally or by design, have spread with astonishing rapidity in their new home. One of the classic instances is that of the Canadian waterweed (*Elodea canadensis*), introduced from North America into England between 1830 and 1840 (the exact date is uncertain). This plant spread so rapidly, both here and later on the continent of Europe, that in a very few years it was a serious pest, blocking ditches and rivers. After a peak of luxuriance, however, it declined, until to-day it is a reasonably common water plant but never a threat to navigation or drainage. The whole of this enormously rapid spread through Europe was accomplished by vegetative reproduction, for almost the entire British and Continental material of this plant is female and hence no seed is set. It has been suggested that a plant clone of this kind may decline through " old age," but there is really no evidence for this view, and many plants we know habitually reproduce vegetatively without loss of vigour. It seems that the real explanation of the decline is as yet unknown. Much has been written about the Canadian waterweed, and the story is well worth reading in more detail (see Arber, 1920, pp. 210-213, and other references given there).

We have already mentioned the common duckweed (*Lemna minor*) as an example of a very widespread aquatic plant. The whole group of plants to which the duckweed belongs (*Lemnaceae*) is of great interest from many points of view. The first and most obvious is because of their small size and unusual form. The layman tends to class duckweed with the green slime, composed of algae, which so often accompanies it in stagnant ponds; but the duckweed *is* a flowering plant in spite of its diminutive size. The duckweed plant (*L. minor*) consists of a single, circular, green, plate-like frond a few millimetres across, which floats on the water, and from the centre of which hangs down a rootlet. In active growth, buds form on the side of the plant and separate off as new plants. Practically the whole spread of the plant is therefore by vegetative budding, which slows up and virtually ceases in the winter, to start again when the warmer season comes. The flowers are extremely small, and are not commonly produced; they should be looked for in shallow water exposed to strong sunlight in a hot season.

Other species of duckweed are more restricted in their occurrence, but may be abundant in some regions. In fenland ditches, for example, the common duckweed is usually mixed with the obviously larger *L. polyrrhiza*, which has a tuft of rootlets. The other floating species which is often locally abundant in lowland areas is *L. gibba*, which has a frond provided beneath with curiously light, swollen tissue filled with air spaces. Both these species produce specialised " winter buds " which contain starch, and lie dormant in the mud at the bottom throughout the winter, to bud off the free-floating summer type in the following season.

In addition to the free-floating duckweeds, a submerged species is very common in many parts of lowland Britain. This is the ivy-leaved duckweed (*L. trisulca*), whose fronds normally remain attached to each other on slender stalks for several generations, thus making a symmetrically branched structure. Flowers are produced only rarely, on the surface, at the edge of special fertile floating fronds which resemble the other duckweeds in possessing stomata.

Finally, in this remarkable group we must mention the smallest flowering plant in Britain—*Wolffia arrhiza*—which appears like green pinheads (usually little over 1 mm. in diameter) on the surface of stagnant water. This plant is rarely recorded in Britain, and, of course, we can say that this is partly due to its insignificance. It probably *is* quite rare, however, because its reputation as our smallest flowering plant ensures that many botanists keep a look-out for it in likely pools.

In parts of south and east England, ponds and ditches are sometimes covered in late summer with a purplish green film composed of the tiny fern *Azolla*. This, like the Canadian waterweed, is a native of North America which has become naturalised in Europe. It belongs to a very peculiar group of free-floating ferns, profoundly modified for their mode of life. Very few ferns are really aquatic. Our only native aquatic species is the pill-wort (*Pilularia globulifera*), whose slender, grass-like fronds betray their fern nature by unrolling at the tip as they develop. The pill-wort is a rather local plant which occurs scattered throughout the British Isles in shallow gravelly or peaty pools; it is hardly to be called rare, for where it does occur it may be in some abundance, forming a sward just beneath the shallow water. The " pills " which give it the common name are curious spherical structures called sporocarps which develop at the base of the fronds; these sporocarps contain spores of two kinds, the female megaspores and the male microspores, and are provided with a most complex wall which is a fascinating object seen under the microscope. The life-history of these water-ferns is complicated and differs a good deal from that of the ordinary ferns. Both *Azolla* and *Pilularia* are rather easy to cultivate in shallow water (the latter rooted in sand or gravel), and are well worth study.

Water plants illustrate very well the phenomenon of similarity in vegetative structure between plants of very different relationships, adapted for life in the same sort of environment. (Every gardener must be familiar with examples of this phenomenon of convergent adaptation, such as the various different plants loosely called " laurels " because they are all shrubs with evergreen " laurel " leaves.) The pill-wort makes a convenient starting-point for a consideration of some of these cases. The stony or gravelly margins of lakes and ponds in upland Britain often show a zone of grass-like vegetation, the individual plants of which bear a group of stiff erect and more or less tubular leaves 2-4 inches long. The pill-wort itself, as we have seen, can readily be distinguished by the unrolling tip of its young leaves; there remain, however, several other possibilities, and as we have seen, aquatic plants tend to flower irregularly so that we cannot usually expect to be provided with floral clues to help us. Perhaps the likeliest plant in this situation, and with this general appearance, is the little relative of the plantains called *Littorella uniflora*. This is distinguished by producing offsets on slender creeping stems, and by its leaves, in section, being flat on one face and round on the other. If the water-level of the pond remains high in summer and the plant is submerged, it makes no attempt to flower; but if in a dry summer a fall in water-level exposes the plant, the whole area will normally flower at once, and the remarkable stamens with slender filaments up to an inch long will be produced.

In many stony northern lakes and ponds the quill-wort (*Isoetes*) may be found around the margins. Like *Littorella*, this plant has a group of long sub-erect basal leaves, which are, however, usually a good deal longer (4-6 inches rather than 1-2 inches) and show, moreover, four tubes in cross-section. The quill-wort is not a flowering plant; it belongs to a small and peculiar group of plants which are usually placed near the ferns, but show no close relationship to any plants alive today. Like the water ferns mentioned above, the quillwort produces spores which are of two kinds, the female-producing megaspores and the male-producing microspores. These are borne in special cavities at the leaf-bases. The quill-wort spreads vegetatively, not by long runners, but by lateral buds which themselves establish young rosettes.

There are three species of *Isoetes* in the British flora, but only two of these are aquatic plants. The common quill-wort (*Isoetes lacustris*) has smooth megaspores, whilst the much rarer *I. echinospora* has, as its name suggests, spiny spores. *Isoetes hystrix*, the remaining British species, is confined to the Channel Islands and the Lizard in Cornwall; it is a curious little plant of peaty and sandy places damp in winter, which produces its leaves in winter and early spring, and survives summer drought as an underground corm. It is, in fact, a typical western Mediterranean " bulbous " plant, adapted for cool damp winters and hot dry summers, and is very different indeed in its requirements from its relatives of northern lakes.

Another plant easily confused, when not in flower, with the quill-wort, is the water lobelia (*Lobelia dortmanna*). It has the same arrangement of long leaves, and frequents very similar lakes and ponds; but the leaves are distinguishable by the possession of a milky juice. They also show two, not four, tubes when cut or broken across. When the water lobelia is flowering, however, which is usually in July or August, there can be no possibility of mistake. Then the delicate two-lipped pale lilac flowers are borne on slender stalks above the water. These flowers are admirably adapted for cross-pollination by insects: but, curiously enough, there seems to be no record of any particular insect visiting the flowers. In Knuth's *Handbook of Flower Pollination* (1906-9) the author reports: " In spite of repeated watching by the Einfelder See (at Neumünster) I failed to observe any [insect visitors]"; and the *Biological Flora* account by Woodhead (1951b) shows that there is still no record of insect visitors. Here is an interesting little problem for further investigation!

These are the three commonest upland lake margin plants with rosettes of narrow leaves. Another might be mentioned, however, for although it is rather rare except in parts of the Hebrides and the Highlands of Scotland, it presents many features of interest. This is a little annual herb, the awlwort (*Subularia aquatica*), the only fully aquatic British member of the large family *Cruciferae* to which such familiar and important plants as cabbages, mustard and wallflowers belong. Its flowers, built on the standard cruciferous pattern with four sepals, four petals and six stamens, are minute, often submerged, and apparently automatically self-pollinated whilst in the bud or only partly open.

It is curious that the Monocotyledons—the group which includes the grasses and many other plants of similar habit—produce very few submerged aquatics with long, grass-like leaves. The few truly aquatic grasses do not have submerged, but rather long loose floating leaves; and even the rushes which may resemble *Littorella* or *Isoetes* in habit do not normally grow in this form when submerged, and none of their species is truly aquatic. Practically the only Monocotyledon which produces a submerged grass-like " sward " in this way is the sedge *Eleocharis acicularis*, but the stems of this plant are so slender and hair-like that it cannot readily be confused with any other plant growing in the same type of situation. Like *Littorella*, this tiny slender sedge only produces its diminutive flowering spikes when the water-level falls and it is left high and dry; and in places where this does not happen, as in the Main Lode at Wicken Fen, where the water-level is kept high in both summer and winter, the plant grows permanently in a vegetative state, and is, therefore, quite unidentifiable in any ordinary Flora. Such plants, if thrown out on mud by the bank during summer dredging, may flower quite freely.

Mention of the awlwort, a solitary aquatic representative of a large family of land plants, raises the interesting general question as to whether this kind of relationship is common and what it implies. We have already seen (p. 148) that *all* aquatic flowering plants must be considered as secondarily adapted to the water, for they share, with normal land plants, flowers which are adapted for aerial and not aquatic development. It is, however, true that some water plants have no near relatives, whilst others obviously belong to familiar and typical land plant families. Let us think of further examples (from the British flora) of isolated water plants in typically land families. (We can often

Plate XIX. Marsh Helleborine, *Epipactis palustris*, with Sea Milkwort, *Glaux maritima*.
Dune slack, Norfolk, August

recognise these in a list because they have Latin names like *palustris* (of marshes) or *aquaticus*—though, of course, this is by no means always the case.) The carrot family, *Umbelliferae*, contains several aquatic species, and in one group in particular, the dropworts (*Oenanthe*), we see a variety of closely-related species ranging from marsh plants (e.g. *O. lachenalii*) to true submerged water plants (*O. fluviatilis* of the Fens)—see p. 177. In such cases the adoption of a submerged aquatic habit (whilst retaining aerial flowers) is obviously to be thought of as a recent happening in evolution. Again, the primrose family (*Primulaceae*) is typically composed of land plants; yet one of our prettiest and best-known water plants, the water violet (*Hottonia palustris*) belongs to it; and, in a family allied to the gentians, there is a remarkable example of mimicry in the so-called " fringed water-lily " (*Nymphoides peltatum*), whose round floating leaves are extraordinarily like those of the true water-lilies, but whose yellow flowers are entirely different. The fringed water-lily, incidentally, is a local rather than a rare plant, occurring in abundance in some sluggish rivers and ditches in the Fens. In all these cases we have examples of *separate* evolution of aquatic species in groups of land plants.

The true water-lilies, however, illustrate a rather different possibility. In this family there are no true land plants, but only aquatic or marsh species—ours, of course, the white (*Nymphaea*) and the yellow (*Nuphar*) water-lilies, are complete water plants. Here it looks as though the group was very old, and that, although some distant ancestral type had, presumably, a land origin (for the flowers are aerial), evolution of the group has gone on in the water ever since and new types have arisen, flourished and been lost, to leave us the present-day water-lilies. In the Monocotyledons, we find other examples—in fact there is a group of Monocotyledons which are wholly aquatic, and to which belong familiar plants such as the pondweeds (*Potamogeton* spp.), the beautiful flowering rush (*Butomus umbellatus*) and the floating frog-bit (*Hydrocharis morsus-ranae*). Some people, indeed (e.g. Prof. G. Henslow), have been so impressed by the number of water plants among the Monocotyledons that they have suggested an aquatic ancestry for the whole group; but there seems to be little real basis for such a sweeping speculation. The whole subject is a puzzling one, and the reader who is interested cannot do better than read

---

Plate XX. Young dunes with Marram grass, *Ammophila arenaria*. Culbin sands, Scotland. (*R. M. Adam*)

Mrs. Arber's excellent book (1920), where this and most other aspects of these plants are dealt with.

We have talked so far, in the main, about *small* free-floating plants; submerged, rooted ones; and a few rooted water plants (such as the water-lily) with floating leaves. It is interesting that we have hardly any examples of *large*, free-floating aquatics in the British flora. This is, perhaps, not surprising, for suitable habitats would not naturally be very common, as the plant would require still and sheltered, but open, water if it were not to drift away. Indeed, there are really only two British water plants which can be said to be reasonably large and free-floating; these are the frog-bit (*Hydrocharis*) and the water soldier (*Stratiotes*). These two Monocotyledons are related, but differ a good deal in general appearance, for the frog-bit has round floating leaves one to two inches in diameter, whilst the water soldier has long, stiff, saw-edged leaves, and looks rather like a floating pineapple-top. The frog-bit is not uncommon in still waters in lowland England, but the water soldier is rather a rare plant, occurring chiefly in the Broads and Fens of eastern England.

The adaptation of water plants to survival through the winter has already been discussed in the case of the duckweeds, and in the frog-bit we find much the same arrangement as in *Lemna polyrrhiza*—the production of special winter buds or turions in autumn which sink and remain dormant at the bottom of the pond until growth begins in spring, when the plantlets rise. In the water soldier, however, a different situation is found. The leaves (like those of a good many water plants) produce an incrustation of carbonate of lime on their surfaces during the growing season when the plant is floating, and this incrustation becomes so heavy by autumn that the whole plant sinks to the bottom. Here it remains in the mud through the winter. In spring new young leaves develop; these are without the lime incrustation, the plant gradually becomes lighter than water and rises to the top again, where active growth and assimilation begin once more. Later young plants are produced on runners from the parent tufts during the growing season and soon detach themselves and live independently.

The flowers of *Stratiotes* are almost always female in this country; and it is said that throughout the northern part of the range of the plant in Europe only, or almost only, female plants occur. It is certainly true that no ripe seed is known in this country, and the spread of the plant must be entirely vegetative. Other cases are known where different sexes of the same plant have different distributions; there is the already quoted and familiar example of *Elodea* (p. 154), but there we are dealing with an introduced plant. One of the most interesting cases in the British flora is that of the butterbur (*Petasites hybridus*), a plant of marshes and river-banks. Professor Valentine has shown (1939, 1946, 1947) that the male plant of this species occurs quite commonly throughout the British Isles, but that the female plant has a curiously restricted distribution, chiefly in the north-west of England —it is, for example, quite common around Manchester. Within this main area the female plants apparently produce abundant seed which in tests has germinated freely to give both male and female plants in the progeny. Valentine suggests that the restriction of the female must be due to some climatic factor which does not operate in the same way on the more adaptable males. He also suggests that some of the wide distribution of the male plant may be due to its having been planted to provide early nectar for bees. Much of the detailed information on this problem has been provided by members of local natural history societies, and, as Valentine suggests in his papers, further information would be welcome.

So far we have dealt with truly aquatic plants; but there are obviously a great many plants which we must characterise as semi-aquatic. The largest group of these are the reed-swamp plants, almost exclusively Monocotyledons with long, narrow, erect leaves, such as the common reed (*Phragmites communis*) and the bulrush or reed-mace (*Typha* spp.). We have already talked about the reed-swamp community in connection with bogs and fens and the hydrosere succession; but it will be interesting at this point to take a few examples of the plants to be found in this zone by a slow-moving river or round the margin of a pond. In various parts of lowland England, usually in ponds, castle moats, and slow-moving rivers, may be found the sweet flag (*Acorus calamus*, Pl. XV, p. 142). As its common name suggests, it resembles the flag or wild iris (*Iris pseudacorus*, Pl. 19, p. 162) in its leaves, which are, however, rather longer, characteristically corrugated, and possess a curiously placed " mid-rib " well over to one side. The easiest way of recognising this plant, however, is by the unmistakable sweet scent of its bruised leaves. It was a favourite in early days for strewing on the floors of churches and baronial halls, and there is no doubt that it was widely spread by man all over western

Europe. Indeed, nowhere in western Europe does it set seed, and we have good reason to believe that in ancient times the plant was brought from Asia. Its sterility in western Europe seems to be due to the fact that the particular type we have is a triploid race with an odd number of chromosome sets (see p. 42)—a cytogenetic sterility rather than (as in the case of *Elodea* and *Stratiotes*) one due to the absence of one sex.

A group of somewhat smaller plants, which would not compete very well with tall reed-swamp, are the water plantains and their relatives (*Alismataceae*). One of these, the local western *Luronium natans*, is indeed, as its name suggests, a true floating-leaved aquatic, but the other common British species of the family normally produce aerial leaves and are to be seen fringing shallow ditches or ponds in lowland England. The arrowhead (*Sagittaria sagittifolia*, Pl. 18*b*, p. 147) shows an enormous plasticity; as we have already mentioned, in swiftly-flowing water, it can grow totally submerged and in a non-flowering state with ribbon leaves; but more normally it produces a crop of characteristic aerial, arrow-shaped leaves as well as submerged ribbon leaves and others of transitional form. The conspicuous flowers have three large white petals; like other members of this family, the flower structure resembles that of the buttercups (*Ranunculus*) closely, but with parts in threes instead of (usually) in fives. The locally common lesser water-plantain (*Baldellia ranunculoides*, Pl. XVI, p. 143) has large pale purplish flowers up to three-quarters of an inch across, which are succeeded by small heads of dry one-sided fruits remarkably similar to those of the buttercups. An interesting variety of this plant, with a creeping and rooting habit of growth, occurs particularly on the margins of lakes in the west of Britain.

The water-plantains themselves (*Alisma* spp.) form an interesting group which has often been treated as a single variable species. It has recently been shown, however, that we have really three distinct species in Britain, two of which are common. There is, first, *Alisma plantago-aquatica*, which normally grows and flowers in shallow water at the edge of a pond or ditch, and produces large, broad aerial leaves. Closely similar to this, and often, in lowland England, occurring with it, is *A. lanceolatum*, with narrower aerial leaves, and somewhat deeper-lilac-coloured flowers; it has also a different fruit-shape. The third species, *A. gramineum*, formerly known only from the shores of an artificial lake near Droitwich, has recently been discovered in quantity

G. Atkinson

Plate 19. Yellow Flags, *Iris pseudacorus.* Kew, May

B. A. Crouch

Plate 20. Grass of Parnassus, *Parnassia palustris,* in flower and fruit. Cumberland, September

on a Lincolnshire river, where it is normally in the submerged state with long ribbon leaves and few or no aerial leaves. The distribution and flowering of these species was investigated by the Dutch botanist, Dr. H. D. Schotsman (1949), who made the intriguing discovery that the three species differed not only in the ways already mentioned, but also in the time of day when the flowers opened. This is an unusual specific difference, but so far as the common species are concerned, seems to be perfectly true both in this country and in Holland. The small flowers are borne on large compound inflorescences, and the petals of each flower last for only part of one day. In *A. lanceolatum* the flowers open quite early (9-10 a.m.) and are beginning to wither by the middle of the day, whilst in *A. plantago-aquatica* they do not open till 1-2 p.m. It seems, therefore, possible, on a summer morning, to pick out the two species simply by noting which plants have freshly-opened flowers. " Opening time " for the little *A. gramineum*, incidentally, is said to be 6.15-7 a.m., but, so far as I have been able to ascertain, no hardy botanist has yet verified this in Britain. Study of the chromosomes of these species has shown that *A. plantago-aquatica* and *A. gramineum* are diploid with fourteen chromosomes, whilst *A. lanceolatum* is a tetraploid with twenty-eight. Such a relationship is known in other species groups, and it seems possible that *A. lanceolatum*—which is intermediate between the other two species—is an allo-polyploid which has arisen from them by hybridisation and subsequent chromosome doubling.

Although they are not perhaps strictly water plants, it would not be out of place to say a little here about a few of the plants which are specially characteristic of stream-sides and river-banks, as they do not readily fall into any wider category. It is interesting that three of our conspicuous stream and river-side plants are fairly recent introductions to this country. These are the tall, purple-flowered Himalayan balsam (*Impatiens glandulifera*), the brown and yellow-flowered water musk (*Mimulus* spp.), and the handsome blue lupin (*Lupinus nootkatensis*, Pl. XVIII, p. 151), confined to river gravels in Scotland. Anyone familiar with the devastated countryside of the industrial north—the South Yorkshire coalfield, for example—will be well aware that the rivers in this area are so polluted that few if any plants can grow in them, or even, in most cases, on their banks. (One plant which does do quite well is the butterbur (*Petasites*), which has already been mentioned in another connection.) In fairly recent years, however, the purple balsam has spread considerably along such polluted rivers and now brightens many desolate and depressing vistas. The plant is annual, and grows in the young seedling stage at a most remarkable rate, producing a brittle, hollow stem and, later, many large, dangling, slipper-shaped flowers. These are succeeded later by " explosive " fruits which when ripe are sensitive to touch and burst scattering their seeds. Little is known about the spread of this balsam along river-banks. It is not even clear why it should be restricted to the riverside habitat, as it is easy to cultivate in ordinary garden soil. Its success along the heavily polluted rivers of some industrial areas is particularly interesting. One possibility, of course, is that it is able to take advantage of the bare, damp ground at the polluted river's edge where other plants are killed, but where its own seedlings, for some reason, thrive. Here is an interesting problem to study—and one with the advantage that it can be studied in places where a field naturalist is not otherwise likely to find very much encouragement!

Other species of balsam occur in Britain, and one, indeed (*I. noli-tangere*), is a native, though rather rare. Two others, *I. parviflora* and *I. capensis*, have spread much in recent years, but mostly in woodland, by riversides, and in waste places in the south of England.

The water musk or monkey-flower is a familiar sight in many western and northern parts of Britain; but the spread of this plant in this country is quite recent (since about 1830). There are two distinct species involved, *M. guttatus*, a native of North America, which has softly hairy flower-stalks and only small red spots on the corolla; and the much less common *M. luteus*, native of Chile, with glabrous flower-stalks and with large red or purplish blotches on the flower.

The study of the plants of rivers, lakes and ponds can provide abundant interest for a field botanist, and, in conclusion, it may be useful to indicate the sort of ways in which a keen amateur could make observations, and perform simple experiments.

Firstly, there is the question of the colonisation of ponds. As suggested earlier (p. 153), a careful study over a period of years of the stages in the colonisation of a newly dug pond or ditch would be interesting and rewarding, as we know remarkably little about it. Secondly, information on the spread of water plants would be very valuable—for example, whether they set good seed and under what condition it will germinate; and observations on the vegetative spread

Plate 21. Felwort, *Gentianella amarella*. Suffolk, August

Plate 22. Sea Lavender, *Limonium vulgare.* Suffolk, July

of species such as *Elodea*. Thirdly, studies on introduced water plants, of which there are several in addition to the ones mentioned, are badly needed. As we well know in the case of *Elodea*, the spread of a successful introduced water plant may be phenomenally rapid. It is probably true to say that the aquatic environment offers the greatest chances of recording vegetational change over a short period, and is therefore a particularly attractive field of inquiry for an amateur naturalist.

CHAPTER 11

THE SEA COAST (J. G.)

B RITAIN, FOR ITS SIZE, possesses one of the most varied and beautiful coast-lines in the world. Some years ago, when flying over the 3,000 monotonous miles of yellow beach between Rio and the Amazon, my thoughts turned to the Dorset coast—to Poole Harbour, Studland Bay, Old Harry, St. Aldhelm's Head and the Chesil Beach—and the full measure of our heritage of coastal beauty was brought home to me.

The great diversity of our coast-line—the sand and shingle of the foreshore, the marram-covered dunes farther inland, the flat salt-marshes at the river's mouth and the sea-sprayed promontory and cliff-face—is matched by a corresponding diversity of vegetation; yet it is a diversity within a unity—a unity provided by the saltness of the sea. In no other habitat is the type of vegetation so obviously linked with a single factor; the majority of coastal plants (with the exception of most dune species, which are sand-lovers, rather than sea-lovers) do not grow naturally away from the influence of salt water, and the presence of maritime species by such isolated inland salt-springs as that at Marcham near Oxford is one of the most striking things in botanical geography.

SAND-DUNES

Imagine you have landed at high tide on a sandy beach some-where in Britain. For the first few yards, there is no sign of vegetable life, but soon, just seaward of the dunes, a thin belt of plants appears among the drifted seaweed. Almost certainly among these early colonists of the beach will be saltwort (*Salsola kali*), sea rocket (*Cakile maritima*), and one or more kinds of orache (*Atriplex* spp.). They are all plants with more or less succulent stems and leaves which store up what little water is occasionally available in this arid strip of foreshore.

Plate XXIa. Sea Pea, *Lathyrus japonicus*, in flower and fruit. Chesil Beach, Dorset, July
(M. C. F. Proctor)

b. Strawberry Clover, *Trifolium fragiferum*. Cambridgeshire, July .*M. C. F. Proctor.*

Plate XXII. Lesser Broomrape, *Orobanche minor*, parasitic on Clover. Cambridgeshire, July

Saltwort, with its spiny recurved leaves, half buried in the sand, cannot be mistaken for any other British plant. Before the introduction of modern methods of chemical manufacture, saltwort, like glasswort (see p. 174), used to be burnt for the production of alkali for glass and soap-making.

The half-dozen or so species of orache (*Atriplex*), which belong to the same family as saltwort (*Chenopodiaceae*), are difficult to distinguish, both from each other, and, at first, from the allied goosefoots (*Chenopodium*); and their somewhat repellent aspect does not add charm to the task. But they have their devotees, and the fact that many of the species like to grow on rubbish dumps and in back yards makes them good spare-time plants for city workers. The oraches have separate male and female flowers, while in the goosefoots the flowers are bisexual; further, the fruits of the oraches are enclosed in two persistent bracteoles which give them a very distinctive look.

The beach species of *Atriplex*, some of them covered with glistening silvery scales, are the least unattractive members of the genus. At least two species (in addition to the more widely spread *A. patula* and *A. hastata*) are concerned, *A. glabriuscula* and *A. laciniata*, but their various forms are still not fully understood—a promising group, in fact, for amateur botanists living by the sea. The name órache is derived, through the French " arroche," from the Latin *Atriplex*.

The sea rocket (*Cakile maritima*), like the saltwort and most of the beach oraches, is an annual, replenished from time to time by fruits washed up among the drift in which it grows, and often shifting its ground on the constantly changing surface of the loose sand. It is a pale mauve-flowered, fleshy crucifer, with a very wide distribution, including Australia and North America. It forms a sort of pair in one's mind with the sea-kale (*Crambe maritima*), another seashore crucifer, but much less common (chiefly found in the west) and apparently decreasing. *Crambe* is a white-flowered perennial, with large roundish glaucous leaves; it prefers shingle, but can occasionally establish itself in sand which is not too frequently disturbed. The leaf-stalks, forced under a flower-pot, are a favourite vegetable in England, but it is very little cultivated on the Continent. It was first popularised in England in the 1790's by William Curtis, founder of the *Botanical Magazine* (see p. 18).

Perennials and biennials do not find it easy to establish themselves in shifting sand, but, besides the sea-kale, there are a handful of others that occasionally succeed. Curiously enough, two of them—the sea beet (*Beta maritima*) and the sea radish (*Raphanus maritimus*)—are also close relations of well-known vegetables. The beet grows all round our coasts, though it thins out northwards, while the radish, with flowers of a darker yellow than the common wild radish (*R. raphanistrum*), is a typical Atlantic species, rare except on the west coast, and confined elsewhere to Holland, Belgium, France, Spain and the Mediterranean.

Two perennial sea species of knotgrass (*Polygonum raii*—named after the great John Ray—and *P. maritimum*) also occur; the first is common, the second exceedingly rare. In anyone unfamiliar with *maritimum*, the first sight of *raii* in a new locality is apt to raise false hopes of the rarer species. *P. maritimum* is confined to a few spots on the south and west coasts. It has a woody stem, glaucous revolute leaves and silvery white ochreae with wavy branched veins; whereas the straggling stems of *raii* are herbaceous, its leaves are flat, and the veins on the ochreae are simple.

Landward of this thin " drift-line " of plants that are just able to keep a foothold in the sand, the dunes begin—at first mere eddies of sand against isolated grass-stems, but farther back growing into the graceful, marram-crested hillocks of Glenluce, Formby, Braunton Burrows, Studland Bay, Blakeney Point and a hundred other such places around the coast of Britain.

Sand-dunes are unique among natural features of the landscape in that they represent a kind of co-operative effort between the vegetable and mineral kingdoms. The full details of this co-operation have been dealt with in another volume in this series (Hepburn, 1952) and I must not go into them here; but broadly speaking, we can say that dunes are formed by blown sand being arrested in its course by plants which, once buried, have the capacity of growing through their covering, both upwards and sideways, thus forming further projections against which further grains of sand can be piled up. The resulting dune, which may reach a height of 100 feet, forms a very specialised type of plant habitat, the main characteristics of which are extreme surface dryness and mobility, and extreme poverty in plant nutrients.

In Britain there are three main plants concerned in dune-formation —all grasses: marram grass (*Ammophila arenaria*, Pl. XX, p. 159), lyme grass (*Elymus arenarius*), and sea couch grass (*Agropyron junceiforme*), though the "drift-line" species already mentioned do sometimes manage to build up little dunelets, which, however, fail to increase, as the plants lack the capacity of pushing through their load of sand. To the land-ward of these drift-line dunelets two phases of the dunes proper can be distinguished—first, the foothills or fore-dunes and, behind them, the main dune range, so full of sheltered picnic pockets on a windy day. On many beaches—for example those near Southport and on the Norfolk coast at Blakeney and elsewhere—the fore-dunes are within reach of the high tides and here they are formed by the sea couch-grass, which, unlike the other two dune-forming grasses, can stand immersion in salt water. These couch fore-dunes are Peter Pans which never grow up, since *Agropyron* has a limited capacity for upward growth through sand and cannot form a dune more than four or five feet high. Its creeping runners, however, can push very deeply into sand or shingle and it is therefore a successful pioneer dune-former, not easily shifted by storms when once established. There is one species—the sea sandwort (*Honkenya peploides*)—which is particularly at home on the couch fore-dunes, and which itself plays a minor role in dune formation. It is an Atlantic species, closely related to the sand-worts (*Arenaria*), but has fleshy leaves, which have been eaten as a pickle. Though plenty of seed is produced, seedlings are very rarely seen (perhaps, as suggested by Sir Edward Salisbury, because the seeds must be scratched by pebbles before they will germinate) and the rapid spread of the plant is largely vegetative. Where the couch-grass is absent the fore-dunes are formed above the high-tide mark by marram, more rarely by lyme-grass, or by both, but it is only the marram which can continue to grow upwards through an ever increasing mound of sand and thus form the main dune range fifty or a hundred feet high.

If you have walked along the crest of a sand-dune range on a hot day there is no need to emphasise the " open " nature of its surface. The plants grow in isolated tufts separated by bare sand which very soon packs itself inside your shoes and makes progress a torture. Farther inland, however, the sand is firmer and the vegetation more continuous. This change is brought about by a number of plants whose rhizomes creep along just below the surface, binding the sand and enabling other species to gain a foothold. These two types of sand-dune—the mobile crests near the sea and the flatter, more stable ground behind them—are known as " yellow " or " white dunes," and " grey dunes," respectively—" white " or " yellow " according to the proportion of broken shells and quartz-grains in the sand, and "grey" because of the lichens which frequently cover the fixed dunes.

In the mobile dunes, between tufts of the ubiquitous marram, a few species grow which are practically confined to this habitat. The most conspicuous are the sea holly (*Eryngium maritimum*), a deceptive umbellifer whose spiny heads of bluish flowers give it the air of a composite or a teasel, the sea spurge (*Euphorbia paralias*) and the sea convolvulus (*Calystegia soldanella*), a beautiful plant with fine pink "trumpets" often resting on the surface of the sand. The sea holly and the convolvulus form a striking contrast in their adaptations to a precarious and unstable habitat. The former, having no creeping rhizomes, cannot bind the sand immediately around it and is therefore liable to alternate exposure and burial. It meets these dangers by firm anchorage, having an extensive root system sometimes reaching to a depth of eight feet, and by a quick response to burial by upward growth through the sand. The rhizomes of the convolvulus, on the other hand, are creeping, and they help to stabilise the sand and minimise its movement by wind.

But the role in stabilisation played by the convolvulus is a very minor one compared with that of the two species chiefly responsible for the development of a grey dune from a yellow or white one. One is a grass (*Festuca rubra* var. *arenaria*) and the other a sedge (*Carex arenaria*). Both have extensively creeping rhizomes sending up tufts of leaves at short intervals. The sedge, especially, by repeated branching, produces an elaborate network of rhizomes which eventually change the once mobile dune into a firm surface on which the marram gradually dies off and other plants, especially certain mosses and lichens, become established.

There are a number of flowering plants characteristic of these stabilised areas—areas over which the narrow pathways along the dunes so often run. Many of them, such as ladies' bedstraw (*Galium verum*) and scarlet pimpernel (*Anagallis arvensis*) are not exclusively maritime. Specially typical of this habitat are the burnet rose (*Rosa spinosissima*), which often covers several acres with a low prickly scrub, and two annuals, the smooth cat's ear, *Hypochaeris glabra* (easily distinguished from the other dandelion-like composites by its yellow florets equalling or falling short of the involucral bracts; see p. 218), and the little mouse-ear chick-weed, *Cerastium semidecandrum* (also easy to identify by the upper bracts being pale and thin in texture). These and other dune-annuals are very characteristic of the habitat. The big, deep-rooted or far-creeping perennials keep their place by a vigorous offensive against their dry and inhospitable surroundings, whereas the little annuals, germinating in the autumn, over-wintering as rosettes, and flowering and fruiting before the summer drought begins, fight an " avoiding action " which, in its way, is equally successful.

Damp hollows in fixed dunes are known as " slacks " and, in western Britain especially, contain a rich and often beautiful flora, including the winter-green *Pyrola rotundifolia*, grass of parnassus (*Parnassia palustris*, Pl. 20, p. 163), felwort (*Gentianella amarella*, Pl. 21, p. 164, and allied forms) and marsh helleborine (*Epipactis palustris*, Pl. XIX, p. 158). The creeping willow (*Salix repens*) is a very frequent inhabitant, forming a miniature forest of yellowish stems a foot or so high, while two species are almost confined to these slacks—the rush *Juncus acutus*, which does not extend farther north than Carnarvon and Yorkshire (and has the sharpest leaves of any British plant), and the narrow-leaved centaury (*Centaurium littorale*), whose southern limit roughly coincides with the northern limit of the *Juncus* (an isolated colony, however, occurs in Hampshire). Narrow-leaved forms of common centaury (*Centaurium erythraea*, Pl. VIa, p. 71), which also grows on dunes, are often mistaken for *C. littorale*. Where their distribution overlaps (e.g. near Southport), the two species occasionally hybridise, but the deep pink flowers, narrow leaves, and long calyx segments, usually rough with minute protuberances, of the true *littorale*, once seen, are never forgotten.

Fixed dunes, in the course of time, gradually develop into a still more stable community, either grassland, heathland, or scrub. The sea-buckthorn (*Hippophaë rhamnoides*), which is so popular in gardens for its orange fruits set against grey foliage, is frequently planted on these old dunes, but in a few areas on the east and south coasts, such as Hemsby in Norfolk, it is a true native and forms an almost pure scrub behind the main dune ridge.

SHINGLE

A shingle beach is clearly very different from a sandy one, not only to the feet of the holiday-maker but also as a habitat for plants. Its stones and pebbles, deposited by the sea in a variety of ways depend-

ing on the direction of the currents in relation to the shore (for example,
as fringing beaches, spits or bars), are too heavy to be blown into
dunes. A long, low ridge like the spit at Hurst Castle, in Hampshire,
or the Chesil Bank in Dorset, dotted here and there with a sparse
vegetation, is a typical shingle landscape. The water-content is much
higher than in sand-dunes, derived mostly from rain, but also from
dew.

Unlike the dunes, with their marram grass, there is no all-pervading
dominant species on shingle. Its vegetation is made up of a number
of plants also typical of other coastal habitats, such as the horned sea
poppy (*Glaucium flavum*), sea sandwort (*Honkenya peploides*), and sea
campion (*Silene maritima*); a miscellaneous collection, especially on
the older and more stable shingles, of inland species, like the ragwort
(*Senecio jacobaea*) and the sow thistles *Sonchus arvensis* and *S. oleraceus*;
and a handful of plants which attain their best development among
the stones and pebbles of a shingle beach. Perhaps the most notable
of these almost " exclusive " species is the shrubby sea-blite (*Suaeda
fruticosa*), practically confined to the east and south coasts, but locally
very abundant where shingle overlies salt marsh as at Blakeney
Point, Norfolk, and on the Chesil Bank, where it forms a long, evergreen fringe beside the lagoon or " fleet " on the inner side of the
beach. *Suaeda* has the power, reminiscent of the marram, of growing
up through shingle that is piled on to it by storms; in fact, the more
it is " shingled over," the more it seems to flourish. Other typical
shingle plants are the sea pea (*Lathyrus japonicus*, Pl. XXIa, p. 166),
the sea lungwort or oyster plant (*Mertensia maritima*) and the dock
Rumex crispus var. *trigranulatus*, often forming a miniature forest with
its dense, brown fruiting spikes; it is a more fleshy plant than the
common curled dock and has a little tubercle on *each* of the three
outer perianth segments instead of on only one.

The sea pea is a lovely plant with bright purple flowers, growing
on a few isolated spots scattered around the British coasts from Ross
to Cornwall. The exposed shingle beach at Aldeburgh, so vividly
described by the botanist-poet Crabbe, is splashed with glowing
patches of it; during a famine the townsfolk are said to have fought
starvation with its unpalatable seeds.

Mertensia maritima, outside Britain, is confined to the bleak coasts of
northern Europe, Asia and America, extending into the Arctic Circle.
It is fairly common in north-west Scotland; its southernmost British

(and European) record was an isolated patch on the shingle of Blakeney Point in Norfolk, now extinct. It belongs to the borage family and like other members, its flowers, set against bluish green, fleshy leaves, are at first red, later blue.

Widespread species, when growing by the sea, are frequently more fleshy than in their inland habitats, though, in many cases, it has not been tested whether these coastal forms are genetically distinct (i.e. ecotypes, see p. 44) or are modifications which would return to normal if grown away from the sea. An example of an untested form is *Solanum dulcamara* var. *marinum*, a prostrate, fleshy type of woody nightshade, not uncommon on shingle along the south coast.

SALT MARSHES

Behind shingle spits, and protected by them, we often find stretches of sandy mud known as salt marshes or saltings. These marshes are formed where sand and mud, brought in by the tide, can be laid down on flat ground which is protected from wave action. The plants of sand-dunes and shingle beaches are not normally reached by sea-water, but the level expanse of a salt marsh is periodically covered by the tide and it is here that a truly maritime or halophytic vegetation is developed—a vegetation that can not only stand, but—in the case of some of its plants—actually requires a periodic soaking in salt water. Most visitors to the sea-side avoid these " dull " stretches of coast, shelterless, windswept and oozy under foot, but to a botanist they are packed with excitement, both for their rare species, and because they illustrate, better than any other habitat, the gradual change of vegetation with a changing environment.

A salt marsh slopes up very gradually inland from the sea, and it is obvious that the seaward portions will be covered by the tide for longer periods over a given time than will those nearer the land. This difference would be comparatively simple if low and high tides were the same throughout the year, but the greater range of the tides at the spring and autumn equinoxes as compared with midsummer and midwinter, introduces an additional complication. V. J. Chapman (see Tansley, 1939, pp. 820-1) has calculated that at Scolt Head, in Norfolk, the lowest zone of the marsh is submerged for 282 out of 732 hours per month during August and September, while the top of the marsh during the same period is covered for only three hours per month. This big variation in length of submersion is reflected in a corresponding zonation of vegetation.

Late summer or early autumn is the season when a salt marsh forgets its normal grey-green monotony and is transformed into a misty purple sheet of sea-lavender. A walk inland from the shore-line early on a fine August morning (there is no shelter and later the sun becomes unbearable) gives a good cross-section of the successive zones and their plant inhabitants.

Nearest the sea, and below the line reached by high water of neap tides, there is a zone of mixed green algae, very often accompanied by grass-wrack (*Zostera* spp.), an ally of the pondweeds (*Potamogeton*) and one of the very few flowering plants that grow actually in the sea. The flowers are enclosed in a sheath near the base of the grass-like leaves. Pollination takes place under water; the curious thread-like pollen is carried by currents to the stigmas which have been sticking out from the sheaths for some time before the pollen is ripe.

Just above the Algae-Zostera zone the otherwise bare, shiny mud is colonised by a sheet of glasswort (*Salicornia* spp.), often covering many acres with its fleshy, leafless stems. Several species and varieties have been described, both annual and perennial, many of them favouring a particular level in the marsh; but discretion in their identification is, for a non-expert, the better part of valour.

In Southampton Water, Poole Harbour, and several other parts of the coast, the glasswort zone is occupied instead by a dense growth of a grass with a remarkable history. About eighty years ago *Spartina maritima* was a rare plant, growing in a few salt marshes scattered round the English coast. Even rarer in Britain was the North American species, *S. alterniflora*, probably introduced into Europe by shipping and first noticed in England in 1829. In 1870 a third *Spartina* was found in Southampton Water. This was described as a new species, *S. townsendii*, and proved to be much more vigorous than the other two. It spread rapidly, and to-day passengers travelling by train from Southampton into Dorset can see hundreds of acres of this grass fringing Southampton Water and Poole Harbour. This *blitzkrieg* by a new-comer is now known to be a striking example of allo-polyploid vigour (see p. 45), as an examination of the chromosomes of the three plants has shown that *S. townsendii* must have arisen as a natural cross between the other two, followed by chromosome doubling.

Plate XXIIIa. The Speedwell *Veronica praecox*, in arable field, Breckland, Suffolk, April

(*M. C. F. Proctor*)

Steps of deserted house, over-grown with Ivy-leaved Toadflax, *Cymbalaria muralis*, and Michael-mas Daisies, *Aster* sp. London, September (*John Markham*)

Plate XXIV. Making a herbarium specimen; showing press, vasculum, mounted lens, and British *Flora*

Here we see, then, the production of a new and successful species of plant within the memory of living men, by a process which no doubt gave rise, during the millions of years of evolutionary time, to many present-day species whose origin is lost in the past. Recent work on *Spartina* has shown that the situation is rather more complicated than was at first thought, because both the sterile hybrid (the product of the original cross) and the fertile allo-polyploid species have spread around our coasts (see Hubbard, 1968).

Another grass, *Puccinellia maritima*, is the characteristic plant of the zone just above the colonies of glasswort or *Spartina townsendii*, though in some marshes (for example at Scolt Head in Norfolk) the sea aster (*Aster tripolium*) takes its place. The purple ray-florets of the aster, framing a yellow centre, normally provide the first touch of colour on a walk from the sea, but there is a form (var. *discoideus*) in which the purple rays are lacking. This ray-less form occurs mainly in England, from the Isle of Man and Whitby (Yorkshire) southwards, though there are one or two isolated records from Scotland. It would be valuable to have a more detailed record of its distribution and to find out whether it is holding its own, or perhaps spreading, in relation to the rayed variety.

Still farther inland sea lavender (*Limonium* spp.) takes up the purple of the aster and spreads it in a continuous carpet over the marsh. There are three salt marsh species. The commonest, and finest, is *L. vulgare*, (Pl. 22, p. 165) though, curiously enough, it does not occur in Ireland. There its place is taken by *L. humile*, closely allied but with less dense spikes of flowers. *L. humile* is fairly common also in England and south Scotland, but the third species, *L. bellidifolium*, is confined to Suffolk and Norfolk. It is a much smaller plant than the other two, and the lower branches of its inflorescence are bare of flowers, giving it a twiggy look which is unmistakable. It is the least spectacular of the three but has an aristocratic distinction of its own. The drifts of its delicate purple flowers in a marsh near Hunstanton will not be easily forgotten by students from the Cambridge Botany School who make an annual pilgrimage there. A second group of sea lavenders inhabit rocky cliffs, and will be mentioned later (see p. 180). A number of other typical salt marsh plants are often found growing in the sea lavender zone. One, *Suaeda maritima*, is an annual relative of the shrubby sea-blite (*S. fruticosa*, p. 172), a dull, unattractive plant until it turns a pleasing red towards the winter. The others are species of sand spurrey, *Spergularia marina* and *S. media* puzzling to distinguish at first acquaintance. Both have fleshy leaves and small pink flowers, but *S. marina* is more prostrate than *S. media*; the seeds of *media* are all surrounded by a broad wing, while in *marina* only a few are winged, the majority having only a thickened border.

At the upper margin of the sea lavender zone the second main colour-giving species, the well-known thrift or sea pink (*Armeria maritima*, Pl. 23*a*) comes into its own. Here the ground is much drier and it is not surprising that thrift is equally common on rocks and sea-cliffs. Its pink flowers first appear in March or April but they linger on to blend with those of the sea lavender in the grand display of August and September. Sheep and cattle often roam over these higher levels in salt marshes, and thrift stands up well to grazing, forming tight rosettes of small leaves, counter-balanced by deep rootstocks. In the more sandy marshes the red fescue grass (*Festuca rubra*) is common in this upper zone and, like thrift, is grazed by sheep and cattle; *Agropyron pungens* (related to the sand-dune *A. junceiforme*, see p. 168) also occurs.

The sea plantain (*Plantago maritima*), frequent in the thrift and sea lavender zones, has become of great interest owing to the intensive study of its many forms by Dr. J. W. Gregor of the Scottish Plant Breeding Station. This variable species occurs over a wide geographic area in Europe, Asia and North America, in a considerable variety of habitats, not only by the sea, in salt marshes and on cliffs, but also on isolated mountains far inland; in some it is grazed and in some not. Dr. Gregor, by cultivating plants from different places and different habitats at Corstorphine, found not only that the species varies over its geographic range, but also that, within a quite small area, different types of habitat, for example exposed rocks and adjoining grassy slopes, carry populations differing slightly from each other in the proportion of the various types of growth-form they contain. It is clear that these populations must have been selected from the original colonising plants in response to the differing conditions of the two habitats (Gregor, 1934-39).

I have so far omitted one plant which is very characteristic of salt marshes but which does not fit conveniently into any particular zone. This is the sea purslane (*Halimione portulacoides*); in old marshes it covers large areas with its grey-green foliage. It is a shrubby species, allied to the goosefoots, and when well established no other plant,

a   *A. Jackson*

b   *A. Jackson*

Plate 23a. Sea Pink, *Armeria maritima*. Kent Estuary, Westmorland, June
b. Biting Stonecrop, *Sedum acre*, on a wall. West Lancashire, July

Plate 24. Meadowsweet, *Filipendula ulmaria.* Inverness-shire, June

except occasionally *Puccinellia maritima*, can survive in competition with it. In younger marshes its favourite home is along the raised edges of drainage channels cut by the scouring tides—Nature's equivalent of ribbon-development.

At its landward edge a salt marsh meets the ordinary non-halophytic vegetation of the neighbourhood. Where the ground is damp, the water dropwort *Oenanthe lachenalii* is a characteristic plant of this marginal zone. There are seven British species of *Oenanthe*, all white-flowered umbellifers and no easier to identify for the beginner than the other members of this family. The best distinguishing characters are in the fruits, but the type of habitat in which you find a species of *Oenanthe* gives a useful first line on its identification. One growing in running water will be *O. fluviatilis*, which has a southern and eastern distribution in England and is very rare outside this country. Its close ally, *O. aquatica*, is the only *Oenanthe* growing in stagnant water, while *O. pimpinelloides* and *O. silaifolia*, alone among the genus, are found in pastures and other grassy places. The remaining three species are marsh-lovers; *O. lachenalii* is commonest in brackish water, though also found in alkaline fens. *O. crocata*, hemlock water dropwort, is virulently poisonous. Being a locally common plant and not unlike celery in leaf and parsnip in root, it has caused many deaths among children. *O. fistulosa*, the smallest species, is also locally common.

Another typical species of brackish ground near salt marshes is *Althaea officinalis*, the marsh mallow. It is a scarce plant, becoming scarcer through drainage, and the common mallow (*Malva sylvestris*) is often mistaken for it. Genera in the mallow family (*Malvaceae*) are rather difficult to separate, but the little bracts at the base of the calyx are a good guide to the three British genera. In *Althaea* (to which the hollyhock, *Althaea rosea*, also belongs) there are six to nine of them, joined together at the base; in *Malva* there are only three, quite separate from each other; while in *Lavatera* (the tree mallow is *L. arborea*, see p. 179) the three bracts are joined at the base as in *Althaea*.

Lastly, two interesting trefoils are typical of grassy places to the landward of salt marshes, *Trifolium squamosum*, and the strawberry clover (*T. fragiferum*, Pl. XXI*b*, p. 166). The former is restricted to our southern coasts, but the latter is common also inland on heavy soils.

CLIFFS AND ROCKS

There could hardly be a greater contrast in plant habitats than between the flat expanse of a salt marsh and the ledges and crannies of a vertical cliff; yet the sea provides a link and there are several halophytic species that flourish in both. Two of them, as we have seen, are thrift and sea plantain; others are *Halimione portulacoides*, *Limonium vulgare* and *Spergularia media*, though other species of the last two genera are more typical cliff plants. Salt marshes are actually covered at intervals by the tides; on a cliff it is only wind-blown spray that reaches the plants, but it is enough to provide the salt they require. Mixed with the true halophytes there are always a number of non-maritime invaders from the hinterland—plants that can stand constant exposure to wind and are tolerant of salt.

The detailed ecology of cliff vegetation has been very little studied, perhaps because the perpendicularity of the habitat has diverted ecologists to more convenient gradients. An adventurous amateur could make very valuable observations during a short summer holiday —for example, a bare list of species with their relative abundance, on almost any cliff round our coasts, would provide information not now available.

There is a handful of plants that grow only on sea cliffs. The first that springs to mind is the sea cabbage (*Brassica oleracea*), the wild species from which the cultivated cabbages, broccolis, and brussels sprouts have been derived. The range of variation in these crops, as compared with the uniformity of the wild species, affords perhaps the best example we have of the power of selection, over a long period, to bring to light, and to fix, the hidden variability of a wild plant. Sea cabbage has a limited European distribution in western France and the Mediterranean, and in this country it is practically confined to the southern and western shores. On the Dorset coast near Swanage the grey Portland limestone forms a perfect background for its lemon-yellow flowers and bluish-green leaves.

Another cliff plant that reminds us of one of our common vegetables

---

Plate 25a. Cowslip, *Primula veris*, and Early Purple Orchid, *Orchis mascula.* Westmorland, April (*B. Garth*).

b. Wild Daffodils, *Narcissus pseudonarcissus.* Westmorland, April (*A. Jackson*).

Plate 25

a

b

C. W. Newberry

C. W. Newberry

is the sea carrot (*Daucus gingidium*). It is connected by a series of intermediate forms with the inland wild carrot (*D. carota*) from which the
garden carrot has been selected. In its extreme form, however, it is
a distinct-looking plant, with fleshy leaves and convex, not concave,
flower-heads. Like the sea cabbage it is an Atlantic species with a
southern and western distribution in Britain.

The plant *par excellence* of sea cliffs is *Crithmum maritimum*, the
samphire of Shakespeare's famous lines from *King Lear*:

> " half way down
> *Hangs one that gathers samphire—dreadful trade!*
> *Methinks he seems no bigger than his head.*"

Its cushions of much-divided leaves fit so snugly into their rock clefts
that they seem to form part of the cliff itself. The gathering of samphire
for pickling has practically died out, but the dish used to be considered
a great delicacy. J. W. White, in his *Bristol Flora*, quoting from the
parish accounts of Weston-super-Mare, states that a peck of samphire
was presented to a seventeenth-century surgeon as a sufficient fee for
setting a broken arm! Two other unrelated fleshy plants, *Inula crithmoides* (*Compositae*) and *Salicornia* species (*Chenopodiaceae*), are also
known as samphire and have been gathered for the same purpose.

Crithmum is an umbellifer and is commonest in southern Britain,
though it does penetrate to Scotland. Another member of the same
family, lovage (*Ligusticum scoticum*) has a reverse distribution. Its dark
green, shiny clumps are abundant on most Scottish cliffs but it does
not get farther south than Northumberland. The leaves have been
eaten as a pot-herb by the Scottish peasants.

I have already mentioned *Lavatera* as one of the three British genera
of the mallow family (*Malvaceae*). *L. arborea*, the tree mallow, is the
handsomest cliff plant we possess. It is often cultivated in gardens and
is an escape in some of its localities. In many, however, it is undoubtedly
native, as in Devon and Cornwall, where its six-foot stems and purple
flowers, nearly two inches across, are a fine spectacle. The name
Lavatera was coined by Tournefort in honour of his friends the brothers
Lavater, Zürich physicians in the early eighteenth century.

From June till August one of the commonest cliff plants in flower
is the sea campion (*Silene maritima*), closely allied to the bladder
campion (*S. vulgaris*) of roadsides. With *Plantago maritima* (p. 176),
these two species share the honour of being the first British plants

to be subjected to a really intensive genetical investigation. The results can be found in another volume in this series, *British Plant Life*, by Dr. W. B. Turrill (1948), who, with Mr. E. M. Marsden-Jones, carried out the investigation over a period of years. Many of the details are highly technical, but some of the broad conclusions catch the imagination. For example, it seems possible that the sea campion survived the ice age in Britain, while the nearly allied *S. vulgaris* is an immigrant to Britain since the close of the Glacial period (see Chap. 4).

Although the salt marsh sea lavender, *Limonium vulgare*, occasionally occurs on cliffs and rocks, another species, *L. binervosum*, and two or three closely allied forms, are more typical of the habitat. They have smaller, narrower leaves, and the flowering stems are branched from about the middle instead of at the top. Three of the allied forms, *L. paradoxum*, *L. transwallianum* and *L. recurvum*, have each been found in only one or two localities, and are not known outside the British Isles, but the reason for their extreme rarity is not known. It is possible that they are recently evolved forms which have not yet spread.

So far I have spoken of a few of the plants that cling to the face of the cliff, " watching," like Tennyson's eagle, from their " mountain walls "; on cliff-ledges a short grassy turf usually develops which also has its characteristic species. In the spring, the little vernal squill (*Scilla verna*) often smothers the turf with a sheet of pale blue—but only on our western cliffs. It is, in fact, a perfect example of an Atlantic species, ranging from the coasts of Norway, through Britain and France to Spain. In the autumn another squill, *S. autumnalis*, occupies similar habitats, but has a much more restricted distribution in Britain, not being found farther north than Gloucestershire. Sir Edward Salisbury has investigated the reproductive capacity (average seed output ×️ average percentage germination) of these two species, together with that of the bluebell (*Endymion (Scilla) nonscriptus*) and finds that the capacity is highest for the commonest species (*E. nonscriptus*), and lowest for the most restricted (*S. autumnalis*), while that of *S. verna* is intermediate—a good example of his general conclusion that one of the factors making for the success of a species is a high seed output coupled with a high percentage germination.

Another Atlantic species on cliffs—and especially on nearly bare, open slopes—is the sea stork's-bill (*Erodium maritimum*). It has a neat, flat rosette of lobed, simple leaves; these distinguish it at once from the other British stork's-bills which all have compound leaves.

Cliff-top swards, especially on limestone soils, are a favourite spot for the little lady's tresses orchid (*Spiranthes spiralis*), with its curiously twisted spikes of fragrant white flowers. Like some other orchids it is very capricious in its appearance. One year the autumn turf may be white with it, while the following year, not a flower can be seen. This extreme fluctuation in numbers is dealt with in another volume in this series (Summerhayes, 1951) and may be due to varying activity of the mycorrhizal fungus associated with the tuber of the orchid.

CHAPTER 12

FIELDS AND ROADSIDES
(J. G.)

" **F**IELD " IS ONE of those conveniently vague words that are useful only so long as you do not try to define them exactly. Practically any more or less enclosed area designed to grow herbaceous plants can be called a field, from a recently planted patch of potatoes to the four-hundred acre Port Meadow near Oxford, continuously grazed since the Domesday Survey of 1085. In this brief chapter I will deal with only two types of field, grass meadows mown for hay and grass pastures grazed by stock (excluding, however, chalk grassland, already dealt with in Chapter 7). Arable fields and their weeds are described in Chapter 13. I have included roadsides in the present chapter partly because they share a certain number of species with fields, and partly because there is no other very obvious chapter to accommodate them.

MEADOWS AND PASTURES

Although a grass field grown for hay is usually called a *meadow*, and one grazed by cattle or sheep is called a *pasture*, there is no really sharp line between them. Fields are frequently grazed in some years and cut in others, while cattle often feed in a field after it has been cut for hay. Broadly speaking, however, meadows and pastures are distinct types of biotic climax community (see p. 40) and each has its characteristic plants. Meadows occur typically on rich alluvial soils in river valleys and are often flooded in the winter, while pastures are usually on drier, more elevated land.

Most of the grasses and other plants found in both meadows and pastures are common species occurring also in other habitats. For

a
C. W. Newberry

b
Eric Hosking

Plate 27a. Scentless Mayweed, *Matricaria maritima* ssp. *inodora.* Derbyshire, August
b. Bindweed, *Couvolvulus arvensis.* Cambridge, August

A. Jackson

b. Field of Buttercups, *Ranunculus* spp. Berkshire, May

F. Reiss

Plate 28a. Poppy, *Papaver rhoeas*, in cornfield

example, the flora of regularly flooded meadows (often called water meadows) largely consists of species found in damp places generally, such as cuckoo flower (*Cardamine pratensis*), meadow-sweet (*Filipendula ulmaria*, Pl. 24, p. 177), marsh marigold (*Caltha palustris*) and many others; while, in drier pastures, many of the common species, such as cowslips and buttercups (Pls. 25*a* and 28*b*, pp. 178, 183) are equally widespread in waste ground, roadsides and similar habitats. But there are a number of plants which are almost confined to meadows and pastures, or, at any rate, particularly characteristic of them. The most famous is the fritillary or snake's head (*Fritillaria meleagris*)— and its most famous locality is Magdalen Meadow at Oxford. Here, in late April or May, the lower-lying parts are thick with its chequered purple bells—a precious possession, generously acknowledged by the authors as one of the few advantages that Oxford can claim over Cambridge.

The fritillary extends, in Europe, from Sweden southwards to the Balkans and, despite its being grown in gardens, there is no reason to suppose that it is not native in the majority of its few scattered English localities, though in some, particularly in the south, it is probably an escape. The Latin name of the fritillary is one of the most picturesque in the British flora; *Fritillaria* is derived from *fritillus*, a dice-box (alluding to the shape of the flower), and *meleagris* is a kind of speckled guinea-fowl (referring to the chequer-board pattern on the petals). *Fritillaria* is a large genus and there are some lovely species recently introduced into rock gardens and alpine houses, in addition to the crown imperial (*F. imperialis*), which has long been a favourite in cottage gardens.

A second very characteristic meadow plant, especially in grassy orchards, is the fragrant yellow tulip, *Tulipa sylvestris*. It is more widespread in Britain than the fritillary, though mainly in the eastern half of the country, but it cannot be considered a native plant. It was the earliest of the wild tulip species to be cultivated in Britain, having been introduced towards the end of the eighteenth century, about two hundred years after the " garden " tulips first reached this country from Turkey via Vienna. Very soon after its introduction it began escaping from gardens and was first found as a wild plant about 1790. It is an interesting species in several ways. Outside Britain it is found over most of Europe and extends eastwards to South Russia and north-west Persia—but its favourite habitat is in vineyards and orchards; hardly ever is it found in a "truly wild" locality. An examination of its chromosomes shows that it is a tetraploid (see p. 45) and it is quite possibly a species which has originated several times and in several places from one or more of the wild diploid species grouped under the name *Tulipa australis*. Its great success in establishing itself in vineyards is probably due to its stoloniferous habit, as it would be easy to introduce a fragment of its stolons with vine-stocks. Even although *Tulipa sylvestris* originated naturally, it would seem that it is now dependent on man-made habitats for its continued existence; it is a good example, therefore, of the difficulty of drawing a sharp line between native and non-native plants (see p. 110).

The two remaining meadow plants that I want to mention are certainly natives—the meadow saffron or autumn crocus (*Colchicum autumnale*), and the wild daffodil (*Narcissus pseudonarcissus*, Pl. 25*b*. p. 178). The *Colchicum*, despite its English names, is not closely allied to the saffron, or to any other form of *Crocus*. It belongs to the lily family (*Liliaceae*), whereas *Crocus* is grouped with *Iris, Gladiolus*, and many other genera in the *Iridaceae*. These two families, together with the daffodil family (*Amaryllidaceae*), are rather easily confused, and it must be admitted that botanists are not fully agreed in which of them to place certain plants, for example the onions (*Allium*). If, however, we follow the classification used in the new *British Flora*, the simplest way to separate the three is to remember that the *Liliaceae* have a *superior* ovary (i.e. the petals come from *below* the ovary), whereas the other two have an *inferior* ovary (i.e. the petals come from the *top* of the ovary). The two families with an inferior ovary can be easily distinguished, as the *Amaryllidaceae* have six stamens and the *Iridaceae* only three. There is no true *Crocus* native in this country, although several garden species are occasionally found naturalised, including an autumn-flowering one, *Crocus nudiflorus* from south-west Europe.

An alternative, and more intriguing, English name for the *Colchicum* is naked ladies—an allusion to the flowers breaking through the ground in August unadorned by stem or leaf. In the following spring the fruit begins to ripen, surrounded by large, glossy-green leaves—an antipodean reversal of the usual northern hemisphere sequence. The plant is locally abundant in damp meadows, and occasionally in woods, over most of England, but is not native in Scotland. An autumn field full of its pale purple flowers is a lovely sight—but not to the eyes of a farmer. It is extremely poisonous in all its parts and has caused many deaths to livestock and to humans. The toxic principle, colchicine, is cumulative in its effect and gradually builds up to lethal concentrations in the body, by which time it is too late to give an emetic. The symptoms, as described by Mr. H. C. Long (1924), are not pleasant and death follows in from sixteen hours to six days after eating the plant. But there is a brighter side to the picture. Colchicine, in suitable doses, is used in medicine for the treatment of gout, and in recent years it has become very important as the chemical by means of which plants can be induced to double the number of their chromosomes.

The wild daffodil is the commonest of the four meadow plants I have selected. Like naked ladies it grows also in woods; but it is at its most spectacular carpeting open fields in solid yellow drifts, as it does, for instance, near Dymock, Newent and many other places in Gloucestershire and Herefordshire. Here the fortunate owners of daffodil fields have commercialised their natural asset and make a charge to the public for picking.

The common *N. pseudonarcissus* possesses what is known, in the terminology of the garden daffodil breeder, as a " bicolor " flower; that is, its central trumpet or corona is of a darker yellow than the surrounding perianth segments. The only other wild British daffodil which may possibly be native has a flower of a uniform brilliant yellow —the Tenby daffodil (*N. obvallaris*). The origin of this plant is something of a mystery. It has been known in pastures near Tenby for more than a hundred years, but has not been recorded outside this country. It may possibly be of garden origin. Mr. E. A. Bowles (1934) describes it as " the most perfect in proportion and texture of any deep yellow trumpet " daffodil—high praise indeed from the most experienced and discerning of all daffodil connoisseurs.

Roadsides

The fact that a plant grows at the side of a road does not seem, at first sight, to tell one much about its ecology, and it is true, of course, that roadsides are very far from uniform in soil or other habitat factors. But there are, nevertheless, certain qualities that a plant must possess to persist in such a locality—qualities that can perhaps best be summed

up in one word, " toughness." Any species that can stand exhaust
fumes from motor cars, the dumping of road-metal and drain-pipes,
and the many other hazards of roadside existence, must have learnt
to look after itself. It is not surprising, therefore, to find that the
vegetation is dominated by plants with abundant powers of
spread and persistence—either by seed or vegetative reproduction
or both.

Among the most successful groups of roadside plants are the
yellow-flowered composites which are lumped together in most people's
minds as dandelions, or—at a slightly more advanced level—as
" hawkweeds, hawksbeards, or hawksomethings." They are certainly
superficially very alike, but, with the exception of the true hawkweeds
(*Hieracium* spp.), they are, like the white umbellifers (see p. 79),
quite easy to distinguish when you get to know them. In Appendix II
(p. 218) there is a list of the species concerned and a key based on
easily-seen characters. Two of the most confusing of the genera involved
are the hawkbits (*Leontodon*) and the cat's-ears (*Hypochaeris*), especially
the species *L. autumnalis* and *H. radicata*, both of which have branched
stems. The most important distinction between the genera is the
presence of scales mixed with the florets in *Hypochaeris* and their
absence in *Leontodon*. This character can be tested quite easily if a
picked flower-head is rubbed vigorously on the palm of the hand so
that the individual florets separate out. In *Leontodon* nothing but florets
will be seen, but in *Hypochaeris* there will also be a number of narrow,
lanceolate, yellowish scales, to the naked eye rather similar to the
florets but easily distinguished under a lens.

The hawkweeds (*Hieracium* spp.) present a very different type of
problem from the other genera concerned. They are largely apomictic
and have evolved a very large number of slightly differing
" forms," many of them growing in the mountainous regions of the
north of Britain, but a few by roadsides. There is a useful summary
of the main groups in the new *British Flora*, and a full monograph by
H. W. Pugsley (1948) which includes 260 forms described as species.
For the whole world, somewhere between ten and twenty thousand
" species " have probably been described!

These yellow composites might all be fairly classed as weeds,
though they are very rarely the same species that are troublesome in
arable land. There are, however, a number of typical roadside
plants which are handsome and distinguished enough to escape this

condemnation. One of the most impressive is also a composite, the woolly-headed thistle (*Cirsium eriophorum*). It reaches a height of five feet, and its solitary, reddish-purple flower-heads, supported by a globular phalanx of bracts interwoven with " spider-web " hairs, are a magnificent sight. It is practically confined to calcareous soils and can be seen at its best on the Oolite lime-stones of the Cotswolds and Northamptonshire. The British plant differs in certain characters from the Continental form and has been separated as subspecies *britannicum*—one of our few endemics (see p. 56).

The teasels (*Dipsacus* spp.) are superficially very like the composites, but they belong to a different family, the *Dipsacaceae*. Their flowers are gathered together in heads, as they are in the *Compositae*, but each flower has a little calyx-like involucre, absent in the composites, and their anthers are not joined together in the typical " tube " of the *Compositae*. There are two British species, both found on roadsides, though the common teasel (*D. fullonum*, subsp. *sylvestris*) is equally at home in open woods and on stream banks, and the second species (*D. pilosus*) frequents dampish, calcareous habitats in general. A second subspecies of *D. fullonum* (subsp. *fullonum*) is the fuller's teasel, which is still grown as a crop in Somerset. Each bract on the receptacle ends in a recurved spine, instead of a straight one as in the common teasel, and the whole fruiting head is used for the dressing of wool cloth. *D. pilosus* is a very different-looking plant from the common teasel and has even been placed in another genus (*Cephalaria*); the white flowers are in spherical, not conical, heads, and the leaves do not join together to form the well-known water-collecting cups of *D. fullonum*. It is a much rarer plant and does not reach farther north than Yorkshire and Lancashire.

The big leaf-rosettes, clothed in soft white hairs, and the handsome yellow spikes of the common mullein or Aaron's rod *Verbascum thapsus*, (Pl. 26*a*, p. 179), lend distinction to many dry roadside verges. *Verbascum* is a big genus, with nearly two hundred species, mostly in the Mediter-ranean region ; in Britain several others occur in addition to the widespread *V. thapsus*, but the only other common species is the dark mullein (*V. nigrum*).

The last roadside species I will mention is the milk vetch (*Astragalus glycyphyllos*), a relative of the purple milk vetch (*A. danicus*, see p. 101), but very unlike it in habit. It is a large, sprawling plant with creamy-yellow flowers and likes best to scramble among the grasses and small bushes of a broad roadside verge. Though local, it grows throughout England and Scotland, but is one of the plants that has not reached Ireland. It is often called liquorice-vetch in books, but has nothing to do with the member of the *Leguminosae* from which liquorice is obtained, *Glycyrrhiza glabra*.

With the increasing mechanisation of agriculture and the large-scale elimination of permanent pasture, roadside verges have become more and more important as refuges for plants (and animals) formerly much more widespread, and their management has therefore come to concern the naturalist. The wholesale spraying of roadsides which was being carried out in the 1950s seems now to be on a much more restricted scale, and an agreement exists between the Ministry of Transport and the Nature Conservancy that such spraying should, in any case, be restricted to main roads or to places where there is some special reason (as on a dangerous bend). Some County Surveyors do not permit spraying on any roadside verge, and in the happiest situations have co-operated with County Naturalists Trusts in discreetly marking special lengths of roadside where there are rare or local species, so that their own workmen will know how they should be treated. Trusts are also co-operating with County Surveyors' departments in experiments on time and frequency of cutting of verges—not an easy problem where the work can no longer be done by men with scythes, but must be mechanised.

CHAPTER 13

ARABLE LAND, WASTE GROUND
AND WALLS (M.W.)

THE ENGLISH COUNTRYSIDE, with its patchwork of fields and woods, demarcated by hedges, is, as we have seen (p. 57), an artificial product of centuries of human activity. Man has created and is constantly changing a whole range of types of habitat, from managed or planted woodland, through permanent pasture, to the transient habitats of arable fields and waste ground. These last are the homes of many plants which we are pleased to call " weeds "; that is, plants which are a nuisance to the farmer and the gardener because they can grow and compete with the plants he is cultivating.

We can hardly think of the rapidly changing populations of plants on arable or waste land as definite communities, each with a characteristic set of species, comparable to a piece of oak woodland, for example; for in these " open " types of habitat, where the plants do not normally cover all the available ground before the next disturbance, there is great variation in kinds and proportions of weeds in different places. Nevertheless it is well known to farmers that certain weeds are successful in certain types of soil and conditions of cultivation; so that, for example, a light acid sandy soil will carry weeds such as spurrey (*Spergula arvensis*) which are absent from a chalky soil. Further, there are weeds common, for example, in the south of England but rare or absent in the north (e.g. scarlet pimpernel, *Anagallis arvensis*); such cases emphasise that climate as well as soil plays its part in determining the weed flora as in determining the more natural one.

In talking of the plants of disturbed ground, we can conveniently begin by considering whether in fact all types of " open " habitat are man-made, or whether some of our " weeds " have not got natural niches where they would survive even without human interference with the vegetation. Sir Edward Salisbury has suggested (1932) that sand-dunes provide an obvious habitat of this kind; and it is interesting to find that some of our commonest weeds, such as chickweed (*Stellaria media*), groundsel (*Senecio vulgaris*) and annual meadow grass (*Poa annua*) occur commonly on dunes. This point has also been mentioned in Chapter 7, where the possible origin of certain weeds from open chalk and limestone cliff and scree habitats was considered. Looked upon from this point of view, indeed, our weeds may be the opportunists who have seized upon man's creation of bare ground and spread from their restricted niches with more or less success. Other plants, as we saw earlier, have apparently not been able to seize the opportunity, and have remained restricted to dunes, cliffs, sea and river strands, etc. In some cases the weed differs from the " native " race sufficiently to be distinguished as a subspecies or species, e.g. scentless mayweed (*Matricaria maritima* subsp. *inodora*, Pl. 27*a*, p. 182) and the native coastal subspecies *maritima*. Direct evidence on these matters is provided by sub-fossil remains (see p. 53).

We cannot, of course, assume that all our weeds are in this category, for in a few cases we know that a plant now common as a weed was not in the British Isles at all until very recent times; a good example is the common annual speedwell *Veronica persica*, which came from Asia Minor and has been in Britain for less than two centuries. In the case of other weeds there is indirect evidence to support the view that there has been a similar history of human introduction.

Baker's work (1947) on the red and white campions provides a particularly interesting example. The red campion (*Silene dioica*) is a woodland and hedgerow plant, obviously at home throughout the British Isles and indeed the whole of north-west Europe. The closely related white campion (*S. alba*) is, however, an arable weed which does not seem to possess a natural home in north-west Europe, but which, nevertheless, in parts of England (e.g. East Anglia) is much commoner than the native red campion. Where the two species meet, they cross, and fertile pink-flowered plants are common. There seems in this case little doubt that the native red campion has suffered from the destruction of the woodland, whilst the white campion probably originally confined to south-east Europe, has spread wherever man has cultivated the soil, until it has reached its climatic northern limit in the north of the Continent.

Plate 29.   Deadly Nightshade, *Atropa belladonna*, in fruit.   Cambridgeshire, August

Stuart Smith

Plate 30. Coltsfoot, *Tussilago farfara.* Cheshire, March

ARABLE LAND

In arable land we find some of our most successful—and there-fore most troublesome—weeds. To be successful a weed of arable land must, it seems, possess one of two main life-forms; it must either be a quick-growing ephemeral (see p. 30), producing abundant, often light, wind-dispersed seed, or it must be a rhizomatous herb, spreading rapidly by vegetative means, and able to grow from any small piece of rhizome broken off in the ground. In the first category come chick-weed, groundsel and shepherd's purse (*Capsella*), each of which may go through several generations in one season; and in the second are the most difficult weeds, such as ground elder (*Aegopodium podagraria*), bindweed (*Convolvulus arvensis*, Pl. 27*b*, p. 182) and couch grass (*Agropyron repens*). It is ironical that in the last class, the *weakness* of the long underground stems is part of the secret of the plant's success as persistent weeds; if when we pulled we could extract the whole plant it would soon cease to be troublesome! The ability to grow from fragments is, of course, very important to the weed; and dande-lions (*Taraxacum*), though tap-rooted, achieve some nuisance value in gardens because of their regenerative powers from root-portions, coupled with their abundant seeding. We are perhaps fortunate that the *Cactaceae* are not hardy plants, for some of these possess astonishing powers of regeneration; in Australia the introduced prickly pear (*Opuntia*) was particularly troublesome, because any fragment of the plant ploughed in is likely to root and grow again!

One of the most interesting aspects of an arable weed flora is the rapidity of change in abundance of many species. In no respect are our nineteenth-century Floras, such as "Bentham and Hooker," more out of date than in their inclusions and omissions of weed species. Flowers such as corn marigold (*Chrysanthemum segetum*), corn cockle (*Agrostemma githago*) and corn-flower (*Centaurea cyanus*), familiar to our ancestors, often described in literature and common up to the middle of the last century, are now rare or local as arable weeds, and many children, even in the country, do not know them by sight. The main change which has eliminated these and many less beautiful and con-spicuous weeds is the purity of agricultural seed. No longer is a high proportion of weed seed sown with the crop; the modern methods of seed purification are too rigorous for most of the old-fashioned adorn-ments of the cornfield. The darnel, *Lolium temulentum*, has changed in

300 years from being a widespread and dangerous grass weed of cereal crops, producing poisonous grain, to a rare casual of rubbish-dumps much sought after by botanists who have never seen it growing! Until very recent years two cornfield weeds persisted in all their glory, the charlock (*Sinapis arvensis*) and the poppy (*Papaver rhoeas*, Pl. 28a, p. 183); these set and scatter abundant seed which can remain dormant in the soil for many years, and are not largely dependent upon being re-sown with the crop. But now, it seems, scientific agriculture has sealed their fate also, for the new development of selective weed-killers, which at the right strength kill most dicotyledonous weeds whilst not harming the cereal crop, seems likely to reduce our corn-fields to an impressive monotony of colour. It is not for the botanist or lover of nature to oppose irrationally such developments in agriculture; but he *might* perhaps be allowed to criticise the growing tendency to use the hormone weed-killers indiscriminately against "weeds" in general. A weed is often, after all, only a plant growing in what happens to be for our purposes " the wrong place "; in the right place it may be a thing of beauty and completely harmless. Thus the whole-sale use of sprays on roadsides to " control weeds " is a costly, mis-guided and muddle-headed attempt to destroy beauty and variety, and substitute uniformity in the name of efficiency and progress. Nothing is gained by such destruction, for the roadside " weed " of grass verges is hardly ever a serious weed of arable land; and often (as we have seen in Chap. 5 and 12) roadsides and hedgerows are refuges for interesting and rare plants which could not conceivably behave as weeds.

In another respect the improvement of agricultural method may well have changed the balance of the weed flora; this is in the wide-spread application of lime and other mineral fertilisers, causing in general an increase in the alkalinity of most arable soils, and the consequent reduction or elimination of some weed species and the favouring of others. It seems likely, for example, that the decline of the annual speedwell with pinkish-white flowers, *Veronica agrestis* (now quite a rare plant in several southern counties of England though locally common in the north of England), which seems to have happened as the introduced *Veronica persica* has spread, is due to its avoidance of basic soils, and not to any sort of competition with the new arrival. In low-land England the blue-flowered *Veronica polita*, smaller and neater than *V. persica*, is quite common also, and does not seem to have changed much in abundance. Again, the corn marigold must owe part of its decline to its calcifuge tendencies; it is certainly still common on acid soils in Norfolk and in parts of Scotland, for example, though in many parts of south and east England it is now rarely, if ever, seen.

A small group of weeds still of some agricultural importance for their damage to crops are parasitic plants, possessing no green chlorophyll themselves, and dependent for their food upon host plants to which they attach themselves. Purer crop seed has, fortunately for the farmer, resulted in a very considerable decline in the dodder (*Cuscuta epithymum*), formerly a serious pest, particularly of clover crops. This interesting total parasite may still be found on heathy ground, where its hosts are often members of the heath family (*Ericaceae*) or gorse (*Ulex*). It appears as slender reddish threadlike stems entwining the host plant, which it may completely kill. There are no leaves, but minute white flowers are produced in clusters on the stems. The seeds are produced abundantly in the autumn, and, though a good deal smaller than those of red clover (*Trifolium pratense*), were formerly an all too common impurity in the farmer's clover seed. H. C. Long (1910) records that 11 per cent of the clover seed samples examined by the consulting botanist of the Royal Agricultural Society in 1900 contained dodder seed, and two samples contained as much as 6 per cent! Other species of dodder are found rarely in this country; *C. epilinum*, which parasitises flax, is occasionally found in crops of flax (*Linum usitatissimum*) from imported seed, but does not persist; and the larger *C. europaea* occurs rarely on nettles (*Urtica*) and the hop (*Humulus*).

The other parasitic plant which is still, in parts of England, a troublesome field weed is the lesser broomrape (*Orobanche minor*, Pl. XXII, p. 167). Broomrapes attach themselves in the seedling stage to the rootlets of their host plant, and produce a thick fleshy flowering stem, often purplish or brownish in colour. The flowers are quite large and have a two-lipped tubular corolla; they are succeeded by capsules containing large numbers of dust-like seeds. It is the abundance, minuteness and persistence of the seed which enables the plant to continue, even with modern agricultural methods, to infest clover crops.

WASTE GROUND

Let us turn now to the plants which are found particularly on waste ground rather than as arable weeds. Some plants, of course, such as the ephemerals groundsel and shepherd's purse, may succeed in either place; but on the whole the characteristic plants are somewhat different. To be successful on the kind of waste ground provided lavishly in and around cities and industrial areas a plant must be an opportunist, producing abundant and effectively dispersed seed. Thus the most striking industrial waste ground weeds, such as coltsfoot (*Tussilago farfara*, Pl. 30, p. 191), rose bay willow herb (*Chamaenerion angustifolium*) and, since the war, the Oxford ragwort (*Senecio squalidus*), produce masses of wind-borne seeds by means of which they reach all waste ground quickly as it becomes available (Pl. 32. The two last examples chosen differ, however, in their degree of stability; the tall perennial rose bay, with its short whitish stolons, spreads rapidly also by vegetative means, and can become the dominant plant on waste ground, particularly industrial slag-heaps, for a period of many years; whilst the short-lived Oxford ragwort, though not strictly an *annual* in this country (as described by Linnaeus), shows no appreciable power of vegetative spread, and is soon crowded out by taller perennial species. For such plants the ability to get seedlings quickly established in newly available, and often most inhospitable, soil is of the greatest importance. One of the most interesting and decorative of the common London bombed site plants is the garden escape *Buddleia davidii*, which was not introduced into this country from China until the end of the last century, but which has in little over fifty years proved remarkably successful in colonising waste ground, where, of course, its shrubby habit and considerable size (up to 15 ft. or so) make it a much more permanent addition than the Oxford ragwort with which it is often to be seen growing. Its long purple inflorescences, produced in the summer, are most attractive to butterflies; the flowers are succeeded by capsules with large numbers of very small seeds.

In addition to the charlock (*Sinapis arvensis*), wild mustard (*S. alba*) and wild radish (*Raphanus raphanistrum*), which are common arable weeds, there are many other crucifers, some very common, others rare casuals introduced with commerce, which occur on waste ground, particularly in towns. Since it is usually very easy to decide that a plant belongs to the crucifer family, with its characteristic four-petalled and six-stamened flower, succeeded by a pod,* it may be worth while to mention here some of the commonest waste-ground

---

* A long pod is called a *siliqua*; a short pod, a *silicula*.

Plate 31. Welsh Poppy, *Meconopsis cambrica.* Westmorland, June

John Markham

Plate 32. Rose Bay Willow-herb, *Chamaenerion angustifolium,* and Oxford Ragwort, *Senecio squalidus,* on bombed site in City of London, July

crucifers and show how they can readily be distinguished from each other.

(A) SMALL ANNUAL PLANTS OF WASTE GROUND (ALSO ON WALLS):

(1) *Erophila verna* (whitlow-grass): tiny over-wintering annual with basal rosette and leafless stem 1-3 in. tall. Small white flowers in spring succeeded by ± elliptic flattened siliculae. Common and variable.

(2) *Arabidopsis thaliana* (thale cress): usually annual, 3 in. to 1 ft., with *simple basal rosette* leaves and few narrow stem-leaves. Flowers small, white, succeeded by long ($\frac{1}{2}$-$\frac{3}{4}$ in.) narrow siliquae. Common on light soils.

(3) *Diplotaxis muralis* (wall rocket): annual (or biennial), usually 6-18 in. Flowers lemon-yellow. Whole plant has characteristic unpleasant smell when crushed, difficult to describe, but to some recalling that of knives being cleaned. Common on waste ground and walls in southern England.

(4) *Cardamine hirsuta* (hairy bitter-cress): common annual weed of arable and waste ground, 3 in. to 1 ft. tall, with a *basal rosette of pinnate leaves* and a few similar stem-leaves. Flowers small, white, succeeded by slender siliquae up to 1 in. long.

(5) *Coronopus* (swine's cress): two species, both *prostrate* annual or biennial weeds with pinnatifid leaves and small white flowers succeeded by *very small (c. 0.1 in.) one- or two-seeded siliculae.* The two species are easily distinguished in fruit; *C. squamatus* has a warty fruit longer than its stalk, while *C. didymus* has a smaller and somewhat less warty fruit shorter than its stalk. Both are common weeds in the south of England.

(6) *Capsella bursa-pastoris* (shepherd's purse): *very* variable; easily recognised by its ± triangular siliculae.

(B) TALLER ANNUAL OR BIENNIAL PLANTS OF WASTE GROUND, ETC.: Several species of *Sisymbrium* are now to be found on bombed sites and other waste ground, in addition to the common hedge mustard (*S. officinale*), distinguished by its straight erect siliquae closely pressed against the inflorescence stem. The commonest is:

S. orientale; tall (up to 3 ft.) branching annual or biennial with quite conspicuous yellow flowers succeeded by long (up to 3 in.) slender siliquae. Its spectacular appearance on London bombed sites during the war led to confusion; for the London rocket of Floras is a different species of *Sisymbrium* (*S. irio*), which spread, apparently in an equally spectacular manner, in London after the Great Fire of 1666, but which is quite rare in this country to-day.

(c) PERENNIAL PLANTS OF WASTE GROUND:

(1) *Cardaria draba* (hoary pepperwort): a recently spread serious weed, common on roadsides and waste ground in many parts of England, with a deep tap-root and producing stolons. Flowers white, on stems 1-2 ft., in May or June, in rather flat-topped inflorescences, succeeded by small siliculae.

(2) *Armoracia rusticana* (horse radish): a common roadside and waste ground plant, originally a garden escape and cultivated for its tap-root. Leaves large, up to 18 in. long, rather like those of docks; flowers white in a much branched spreading inflorescence. *Fruit hardly ever formed in this country.*

(3) *Rorippa silvestris* (creeping yellow-cress): usually in wet places, but occasionally a garden weed or on roadsides and waste ground in quantity. Spreads by stolons. Lower leaves pinnate. Flowers bright yellow, c. $\frac{1}{4}$ in. across, succeeded by siliquae c. $\frac{1}{2}$ in. long.

(4) *Barbarea vulgaris* (winter-cress or yellow rocket): also usually occurs in wet places, but may be found on roadsides. It has a rather stout erect branching stem with a conspicuous inflorescence of bright yellow flowers, succeeded by siliquae.

Quarries, particularly in chalk and limestone, are usually the home of a rich variety of plants, many of them rare or local. The famous deadly nightshade (*Atropa belladonna*, Pl. 29, p. 190), for example, is more often to be found in abandoned chalk pits than in its more natural scrub or woodland habitats. Its much commoner relative, woody nightshade (*Solanum dulcamara*), is often misnamed 'deadly'; but the berries of *Solanum*, though poisonous, are not dangerously so in small doses.

WALLS

Old walls, particularly in the milder and damper parts of Britain, often have a rich growth of vegetation, and certain flowering plants, such as species of stonecrop (*Sedum*, Pl. 23*b*, p. 176), ferns and mosses are particularly associated with them. As we would expect, the natural home of such plants is almost always on cliffs or rocky slopes (see p. 110) and they have taken advantage of these new homes which man has provided for them. In the extreme south-west of Britain, or in Ireland, the climate is so mild and damp that almost any wall will in a relatively short space of years be covered, first by lichens and mosses, then, as the plant debris (which soon becomes peat) accumulates, by ferns and flowering plants. Old field walls in Cornwall, for example, are often difficult to see at all for the covering of growing plants over the buried stones. In towns, of course, the process is not usually allowed to go so far; but it is surprising how many different kinds of plants may be found by careful examination growing rooted in cracks of masonry. Rishbeth (1948) found 186 different plants during a survey of Cambridge walls.

Some of our best-known wall plants are true natives of this country. A good example is pellitory-of-the-wall (*Parietaria diffusa*), which also occurs in abundance on sea-cliffs, and these are undoubtedly its natural haunts. The curious pennywort or navelwort (*Umbilicus rupestris*), so common on walls in the west of Britain, is similarly a plant of natural rock habitats. The little spleenwort ferns (*Asplenium* spp.), which take advantage of man's provision of rock crevice habitats, can also be found on natural rock.

Others, however, and among them one of the best-known wall-plants, the ivy-leaved toadflax (*Cymbalaria muralis*, Pl. XXIII*b*, p. 174), are introductions into this country. This toadflax is a very pretty plant, with delicate trailing stems and two-lipped and spurred lilac flowers; it was first recorded in this country in 1640, but now occurs on old walls almost throughout the British Isles. The capsules containing the seeds are borne on long stalks which react to light in such a way as to put the capsule into dark cracks, where the seeds are shed and can germinate to give new plants. This species is related to the handsome yellow toadflax (*Linaria vulgaris*, Pl. 26*b*, p. 179), common on roadsides and waste ground.

The tower-cress (*Arabis turrita*), a plant of rather shady limestone rocks in south Europe, has had an interesting history in this country. At one time, in the eighteenth century, it was apparently quite common on old walls in both Oxford and Cambridge, though we do not know exactly how it was originally introduced there; but its old walls were pulled down or repaired, and its stations dwindled, until it seems as though there is only one remaining to-day. It is interesting to contrast this story with that of that other " university wall plant," the Oxford ragwort (*Senecio squalidus*—see p. 194), which has achieved such an astonishing recent spread after being content for centuries (it was in Oxford when Linnaeus visited the Botanic Garden) with a restrained existence on a few old walls, docksides, etc. Why is one plant successful and another not? We do not know the answer to these questions; but we can at least say that the light, air-borne seeds of the ragwort will carry it from wall to wall or waste ground far more easily than the rather large round seeds of the tower-cress.

The wallflower (*Cheiranthus cheiri*) has been so long grown in gardens, now in many colour-varieties, that it is difficult to decide where it is really a native plant. In Floras, it has always been assumed to be introduced, because of this long association with man; but anyone who has seen the " wild " yellow wallflower growing on the rocks in the Avon Gorge at Bristol, for example, may be forgiven for harbouring a doubt in their minds as to whether this, at least, is not a native station. Be that as it may, the " wild " wallflower brightens many a ruined building and flourishes without the aid of man to-day.

The yellow welsh poppy (*Meconopsis cambrica*, Pl. 31, p. 194), familiar in cottage gardens, presents a similiar problem in many of its wild localities, but botanists are generally agreed that the plant is native on shady rocks in a number of western counties from Devon to Carnarvon.

We should not close this chapter without a reference to what Lousley has recently called " the bombardment of alien plants." There is a constant accidental and intentional introduction into this country of seeds and fragments of living plants from all over the world, and Botanic Gardens, waste ground at docks, corn mills and many other points of entry are continually allowing the escape of new alien species of plants. Usually such plants, if they are lucky enough to establish themselves for a season or two, soon disappear; they are often frost-sensitive and adapted to milder climates than ours. But just occasionally one settles down, and, even more rarely, one begins to spread. We have mentioned some instances of successful aliens already (see pp. 154, 190); but for every one that succeeds, and wins a permanent place in the flora, a hundred fail. It is, however, interesting and quite important to try to keep some check on arrivals, including the hundred which fail, if only so that we shall have the early history of the rare one when it succeeds! That is why the ugly waste ground near a large port, big brewery, or corn mill attracts some field botanists; for such places provide a bewildering selection of weeds from other parts of the world, and challenging problems in identification. So if you think you know the British flora—a surprisingly difficult goal to achieve—remember that it is not, and cannot be, a fixed thing, but is continually receiving, and occasionally admitting into full membership, a large influx of " foreigners." Strictly speaking, then, the only way to know the British flora is to know that of the whole world!

Anyone who would like to learn more about our weed flora is strongly recommended to get Sir Edward Salisbury's book in this series, *Weeds and Aliens* (1961). Sir Edward has personally studied in some detail many problems of weed floras, such as the efficiency of seed production, dispersal and germination, and his book contains references to most of the important studies of general botanical interest.

CHAPTER 14

HOW TO GO FURTHER (M.W.)

IF THE AUTHORS of this book have succeeded in their design, the
reader should have been left with the desire to find out more about
the flowering plants of our islands. There are several ways of doing
this; but one thing is basic and indispensable to all of them—to go
out in the field and learn by experience. Books, lectures and illustrations
help, but they only supplement and bind together the things one learns
at first hand. An intriguing fact is that one does not recognise, say,
a buttercup by noting the floral characters of " 5 nectary-bearing
petals and an ovary consisting of numerous free carpels "; nor does
the expert consciously " identify " the plants he knows in the field,
although, if taxed, he may well be able to produce the technical
characters to prove his identification correct. This is obvious enough,
of course, in the " identifications " we all make of Mr. Smith or Mrs.
Jones—indeed, although we could recognise Mr. Smith at fifty yards,
we would often be quite unable to describe him accurately to a
stranger!

For many field botanists, this gradual process of getting to
know plants intimately is sufficiently absorbing and fascinating, and
botanical science has been and always will be dependent upon the
existence of sharp-eyed and enthusiastic amateur naturalists for its
basic knowledge of what plants grow where. But there are at least
three other distinct though interconnected directions in which the
amateur botanist can branch out, with great interest to himself, and
with value to botanical science as a whole.

THE STUDY OF CRITICAL GROUPS

The first, which is a logical extension of an interest in the identification of plants, is in the direction of the more detailed and critical study

of the so-called "difficult" genera or species-groups, using the "traditional" methods of intensive study in the field and the herbarium, and of a critical survey of the literature published about the group. There is still a great deal of work to be done on the British flora along these lines, especially in linking up researches on our own flora with recent work on similar or allied plants in other countries. Recent literature from Continental countries whose floras are most likely to resemble ours (i.e. France, Holland, Belgium, Germany and Scandinavia) is a very fertile source of inspiration for British botanists. The abstracts of foreign papers given in the B.S.B.I. journal *Watsonia* are particularly useful in this connection. And the traffic is, of course, two-way; in the north of France, for example, even a short holiday visit may enlarge the knowledge of the distribution and characters of British forms as yet inadequately recognised or recorded on the Continent.

For those who read French and German, indispensable reference works are available in the Abbé Coste's three-volume *Flore de France*, and Hegi's monumental *Flora von Mittel-Europa*. Both are from time to time available; a new edition of Hegi is now appearing; but second-hand or new, they are both expensive, and, like " Sowerby," a luxury to be worked for by the keen botanist who is building up his reference library. In Scandinavia, the *Flora* of J. Lid (1963) is outstandingly good for those with patience to acquire sufficient basic Scandinavian to tackle the Norwegian text.

The first two volumes of a comprehensive *Flora Europaea* (Tutin, T. G. et al. 1964, 1968), in English, and covering the whole of Europe from Iceland to the Ural Mountains, are now published. It is planned to complete this work in five volumes, and there are subsidiary plans to prepare both illustrations and distribution maps of the species included.

Amateur botanists have always played a big part in this detailed study of critical groups and are continuing to do so. Among recent researches, those of the late H. W. Pugsley on *Euphrasia* and *Hieracium* (1930 and 1948), of J. E. Lousley on *Rumex* (1939 and 1944), and of D. P. Young on *Epipactis* (1952), come to mind, and many others could be quoted.

Ecology

The second main field which is open to an amateur who has acquired a good working knowledge of the flora is that of ecology, and particularly the study of single species (*autecology*). The knowledge of where a particular plant grows is naturally followed by the desire to know *why* it grows there, and not elsewhere. This question can, of course, be answered on very different levels. We may, for instance, begin by making the loose and unscientific statement that it grows there because it " likes " chalky soils or acid bogs or whatever it may be; but we cannot leave it at that. Is the plant *strictly* confined to such soils? To answer this question we must see the plant growing, and make notes on the aspect, the soil, and the other species growing with it, in as many different habitats as possible. If it *does* prove to be strictly confined to chalky soils, we might then ask " Will the plant actually *die* if I try to grow it in a different soil in a garden? " We shall then be involved in simple cultivation experiments with the plant of our choice. Further, we might ask: Is the plant able to " hold its own " where it grows; does it flower freely and reproduce from seed, spread vegetatively, or both; how is it pollinated; is it eaten by animals; what diseases attack it?—and a host of other questions. It is surprising how little information of this kind has been collected, let alone published, for British plants; and it was in an attempt to remedy this deficiency that the British Ecological Society began, in 1938, its scheme for a *Biological Flora of the British Isles*. Anyone who is interested in this type of autecological study could not do better than to read the introduction on the aims of the *Biological Flora* and look at some of the numerous instalments already published.

Ecology includes all studies of the relationship of plants to their environment, and must deal with problems of the inter-relations of plants, soil and climate in complicated types of vegetation such as a mature woodland, as well as the superficially simpler problems of the autecology of individual species. But clearly the two are interconnected, and indeed it might be claimed that it is only through detailed autecological studies that the broader problems can be solved. The British Ecological Society, containing both plant and animal ecologists, amateur and professional, caters for the worker in this field; like the B.S.B.I., it holds meetings, exhibitions and excursions, and publishes a journal (*The Journal of Ecology*).

It is not, however, simply the cause of " pure " science which is served by ecological study on a scale possible to an amateur naturalist. Recently, the gaps in our knowledge of the autecology of species have been coming to the fore in another connection. In the first chapter we mentioned the importance of the conservation of the British flora. In an over-populated and highly industrialised island there is a real danger that in the interests of a short-sighted policy of economic gain or other national interest we shall gradually lose more and more of our native vegetation and its great variety of wild plants. The creation by Parliament of the Nature Conservancy in 1949 is a striking recognition of the claims of nature conservation, for both scientific and educational reasons.

The conservation of vegetation and its component plants is, however, no simple task, and is only beginning when the obvious man-made threats of " development " are dealt with. It is then that we realise the complexity of the problem. For example, a fen, if protected and left alone, will soon cease to be a fen at all, and by natural processes of development, will become woodland; if we do nothing more, we start by protecting a fen and finish by losing it! The understanding of these complex processes of development and change is of fundamental importance in this problem of nature conservation. Indeed, unenlightened conservation may well threaten or destroy the very feature it is designed to protect—by preventing the sort of interference (by grazing animals or man) which the particular plants may need for their survival. For the majority of our rare or local species, it is probably true to say that *some* interference with their habitat is necessary for the plant's survival, or at least for its persistence in any quantity. It is when we are faced with the problem of *exactly* what degree of interference is needed that our lack of knowledge of autecology becomes painfully apparent. It may seem a far cry from the work of an amateur naturalist in observing, say, the germination of seeds of the pasque flower in its native haunts, to the functioning of a government department charged with nature conservation; but the success of the latter is in fact dependent to some extent on the information provided by the naturalist!

EXPERIMENTAL TAXONOMY

The third possible line of development comes under the general heading of experimental taxonomy. This comparatively recently developed field has already been discussed in Chapter 2 and is, in origin, a synthesis of taxonomy, ecology, genetics and cytology (see Chap. 3). It attempts to investigate and delimit the units which are involved in the actual process of small-scale evolutionary change. In general, it is true to say that work on experimental taxonomy can be fully carried out only by someone with facilities for examining chromosomes and for growing and breeding plants on a fairly large scale. Nevertheless, the contributions of the amateur botanist's careful records can be extremely valuable, as the B.S.B.I. has recently been able to test in its "Network Research" projects, in which populations of selected common plants are studied in detail by members of the Society and their results analysed by specialists. An account of the first two projects, on comfrey (*Symphytum*) and the bladder campions (*Silene vulgaris* and *S. maritima*) can be found in Perring and Savidge (1968).

I have indicated in this final chapter some of the directions in which amateur botanists can proceed beyond the comparatively elementary knowledge of the British flora outlined in the earlier part of the book. We hope that some, at any rate, will follow up the hints that we have given and graduate, so to speak, to the more advanced stage represented by Turrill's volume in the same series, *British Plant Life* (1948). In it will be found a summary of experimental taxonomic work on the British flora and, in addition, stimulus and inspiration to fill in the many gaps that still exist.

WAYS AND MEANS (J. G.)

IN THIS APPENDIX are given (1) brief particulars of the equipment necessary for the collecting of specimens and the making of a herbarium, and (2) notes on some of the books covering various aspects of field botany in Britain.

COLLECTING SPECIMENS AND MAKING A HERBARIUM

For collecting in the field the absolute essentials are a hand-lens (magnification between x 6 and x 10), a note-book and pencil, a knife for cutting tough stems, a map marked with the National Grid, and either a vasculum (a metal collecting case with a shoulder strap) or a portable flower-press. Small tie-on labels are also useful to distinguish similar specimens which may be confused on returning home. The great majority of collectors prefer a vasculum to a press, especially if the plants are to be examined and identified later. When collecting herbarium specimens of plants already known to the collector, a press has many advantages, although it is awkward to handle in the rain. The National Grid map is important because it is becoming increasingly the practice to give the grid-reference in addition to a place-name, when recording the locality for a plant.

When a specimen brought home in a vasculum has been sufficiently examined (see Pl. XXIV, p. 175), it should be pressed as soon as possible. A plant press consists of a pile of sheets of absorbent paper with wooden or metal " slatted boards " at the top and bottom, and pressure is applied either by straps, or weights on top, or both. A press and special drying paper can be bought from one of the several firms supplying naturalists' equipment, or can easily be put together at home. Newspaper is suitable; blotting paper is not. Specimens should be laid on the paper with great care, so that leaves and petals are flat and as much of the plant as possible is displayed. Unless a species is rare, the whole plant, including roots, should be collected and pressed, and fruit as well as flowers should be present. It is essential to insert a rough label with each specimen giving its name, date and place of collection, and other particulars that may be required for special purposes. Specimens are often difficult to lay out flat without creases, and it is a good plan to tackle them again a few hours after insertion, by which time the pressure will have made them more tractable. Thick roots may have to be sliced in half before pressing, and very large specimens (e.g. the giant Heracleums) may demand three or even more sheets for their accommodation. There is no hard-and-fast rule about frequency of paper-changing, nor about total time in the press, as so much depends on the initial succulence of the specimen and the state of the atmosphere. Normally the paper should be changed every two to three days until the specimen is quite dry. This can be tested by placing it against your cheek and if it feels at all cold or damp, further drying is still needed. It is an excellent plan to put the press in an airing cupboard or in the sun, if either happens to be available, as this shortens the drying time very considerably.

When a specimen is really dry it may be mounted at once or placed loose (with its label) between sheets of thin paper until time can be found for final mounting. The size and type of mounting paper used vary to some extent, but a good, strong sheet of cartridge paper about 17 in. x 11 in. is suitable. Mounting methods are a subject of controversy. The big public herbaria often paint glue on one side of the specimen and stick it to the sheet. This is quick and permanent, but many private collectors prefer to use strips of sticky paper; heavy specimens can be fixed with adhesive linen tape or sewn with thread.

Herbarium specimens are very liable to attack by insects if not properly protected. Here again, several methods are used. In some big collections the sheets are poisoned by passing them through a cyanide chamber or painting the specimens with a solution of phenol and mercuric chloride in alcohol, but, for a private herbarium, moth-balls should be sufficient to keep away intruders. If an attack takes place, the specimens affected can be treated with the mercuric chloride solution mentioned above. Fungus pests are, of course, also a danger if the sheets are not kept in a dry atmosphere, but storage in an ordinary, regularly used room should be safe enough.

Once the specimens are mounted on their sheets they should be labelled, placed in folded paper covers, and stored in a wooden or metal cabinet. Suitable labels can be bought, but many collectors prefer to have them specially printed with their own headings. The minimum essential information on each label should include a collector's reference number for each gathering, the name of the species, the date and place of collection (including vice-county and grid reference), and the name of the collector, but it is also useful to give particulars of the habitat (including height above sea-level), the names of the chief accompanying species, and the frequency of the plant in the spot concerned; other information may be added for special purposes, for example any insect visitors or fungus diseases noted.

The arrangement of the sheets in their covers and in the cabinet depends on the nature and extent of the herbarium. Normally, stout paper covers for each genus, with thinner covers for the sheets of each species, are sufficient, but further subdivisions may be desirable in larger herbaria. The sequence of the families, genera and species should follow the order of some standard work.

BOOKS

British field botanists, there is no doubt, are extremely well served with books and periodicals to aid them in their work—perhaps too well served, as regards quantity, in certain departments, notably popular books on common wild flowers! For many years, it is true, the absence of an up-to-date and comprehensive British Flora was a national disgrace in botanical circles, but the appearance of the *Flora of the British Isles* (Clapham, Tutin and Warburg, 1952, Ed. 2, 1962) happily reinstated us in the comity of nations.

The books, excluding general botanical text books and periodicals, useful for a field botanist up to the stage of "medium expert" can be grouped under the following heads:

Identification (Elementary)
Identification (Advanced)
Illustrations
Plant Lists and Distribution
Local Floras
Books on Special Aspects of the Flora
History and Bibliography

Plant Names and Botanical Terms
General Ecology
Systematic Botany, Experimental Taxonomy and Evolution

At the end of this Appendix a list is given of a selection of books under the above heads. It has not, of course, been possible to include anything like all the books that are available, and the omission of a particular volume does not necessarily mean that it is not considered as good as others that are mentioned. It would clearly be invidious to comment in detail on the books in this list, but some general remarks may, perhaps, be useful.

There is a very wide choice of beginners' books for identification, but we would strongly urge the acquisition of Clapham, Tutin and Warburg (1962) at as early a stage as possible, or at any rate their *Excursion Flora* (1959). Provided you have mastered the necessary technical terms these Floras should present no great difficulty, and you will be learning the accepted Latin names and full distribution of each plant, thus saving the unlearning of much out-of-date information. A complete series of illustrations to accompany these Floras is available; this, together with other series already published, in course of publication, or planned, will provide a great wealth of pictures for British field botanists.

A local Flora exists for each of the great majority of British counties, and the appropriate book for your area should be one of the first to be bought. Lists of local Floras are given in Druce (1933), and in Turrill (1948). Simpson (1960) is a most useful Bibliography of the British flora, and there is an up-to-date list of available local Floras in Perring (1969), which gives details of price, publisher, etc.

Popular books on wild flowers are legion. Some botanists dislike them as a class, while others keep a pile of them on their bedside table. It would be presumptuous to comment on the divergent temperaments thus exemplified.

In the realm of ecology, besides a number of good introductions to practical work, Britain is fortunate in having one of the great classics of descriptive ecology in Prof. Sir Arthur Tansley's monumental *The British Islands and their Vegetation*; this will tell you all you want to know about the plant communities of Britain.

As we have pointed out elsewhere, the history of British field botany remains to be written, and much of the information available in print lies buried in obscure periodicals; but good outlines of general botanical history and bibliography, in Britain and elsewhere, can be found in the books listed under this heading at the end of the chapter.

In general, we would strongly urge field botanists to concentrate on building up as good a library as they can afford. Apart from its usefulness, they may well find that they are implanting a rival, but compatible, passion in their breasts, and that the collecting of books may come to share a twin throne in their affections with the collecting of plants.

A SELECTION OF BOOKS USEFUL TO BRITISH FIELD BOTANISTS

(Full references are given in the Bibliography)

Identification (Elementary)

Bonnier, 1917 (1952); Johns, 1949; Keble Martin, 1965; Lewis, 1958; McClintock and Fitter, 1956 (1965); Makins, 1939 (1952); Meikle, 1958; Prime and Deacock, 1948 and 1970; Step, 1940 (1951).

Identification (Advanced)
(Botanical Monographs on small groups are not included)

Babington, 1922; Bentham and Hooker, 1924 (1947); Butcher, 1961; Clapham, Tutin and Warburg, 1959 and 1962; Druce, 1930 (1943); Gilbert-Carter, 1936; Hubbard, 1968; Jermy and Tutin, 1968; Moss, 1914 and 1920; Syme, 1863-92.

Illustrations with little or no Descriptive Text

Clapham, Tutin and Warburg, 1957 and 1965; Butcher and Strudwick, 1930 (1946); Fitch and Smith, 1924 (1949); Ross-Craig, 1948—.

Plant Lists and Distribution

Dandy, 1958; Druce, 1928 and 1932; Mathews, 1955; Perring and Sell, 1968; Perring and Walters, 1962; Watson, 1883.

Books on Special Aspects of the British Flora

Avebury, 1905 (Life Histories); Edlin, 1951 (Uses); Godwin, 1956 (History
   of British Flora); Hepburn, 1952 (Sea Coast); Lousley, 1950 (Chalk
   and Limestone); Pearsall, 1950 (Mountains and Moorlands);
   Pennington, 1969 (History of British Flora); Raven and Walters, 1956
   (Mountain Flora); Salisbury, 1942 (Reproduction); Salisbury, 1952
   (Downs and Dunes); Salisbury, 1961 (Weeds and Aliens); Summerhayes,
   1951 (Orchids); Turrill, 1948 (Biology, Distribution, Ecology, Genetics,
   etc.).

History

Britten and Boulger, 1931; Gilmour, 1944 (1946); Green, 1914; Gunther,
   1922; Jackson, 1881; Oliver, 1913; Raven, 1947 and 1950.

There are also a number of biographies of individual botanists; see
Gilmour, 1944 (1946).

Dictionaries of Plant Names and Botanical Terms

Britten and Holland, 1878-9; Gilbert-Carter, 1964; Jackson, 1928 (1949).

General Ecology

Ashby, 1961 (1969); Maclean and Cook, 1946 (1950); Tansley, 1939 (1953);
   Tansley, 1946 (1949); Tansley and Price Evans, 1946.

Systematic Botany, Experimental Taxonomy, and Evolution

Briggs and Walters, 1969; Darlington, 1963; Davis and Heywood, 1963;
   Heslop-Harrison, 1953; Heywood, 1967; Jeffrey, 1968; Lawrence, 1951;
   Willis, 1966.

KEYS
(J. G. & M. W.)

It is strongly recommended that the basic botanical characters of the genera and species concerned (see Clapham, Tutin and Warburg, 1952) should be learnt as soon as possible, but it is hoped that the following keys may be of use to beginners in finding their way among these initially puzzling groups.

KEY TO THE COMMON HEDGEROW AND ROADSIDE UMBELLIFERS

1	Flowers yellow or yellowish.	2
	Flowers white or pinkish.	5
2	Leaves v. much divided, segments hair-like, not all in one plane.	*Foeniculum vulgare*
	Leaves 1 or more-pinnate or ternate, leaflets flat, all in one plane.	3
3	Leaves 1-pinnate.	*Pastinaca sativa*
	Leaves 2-3-pinnate or ternate.	4
4	Leaves 2-3-pinnate, leaflets finely serrate.	*Silaum silaus*
	Leaves 3-ternate, dark green and shiny, leaflets obtusely serrate or lobed.	*Smyrnium olusatrum*
5	Stems spotted or blotched.	6
	Stems unspotted.	7
6	Stems glabrous.	*Conium maculatum*
	Stems hairy.	*Chaerophyllum temulum*

7　Bracts 0-1.[1]　　　　　　　　　　　　　　　　　　　　　　　　8

　　Bracts 2 or more.　　　　　　　　　　　　　　　　　　　　　21

8　Plant decumbent, with sessile or sub-sessile umbels opposite the
　　　leaves.　　　　　　　　　　　　　　　　　*Torilis nodosa*
　　Plant erect, with stalked umbels.　　　　　　　　　　　　　　9

9　Bracteoles 0 (rarely 1 or 2).　　　　　　　　　　　　　　　10
　　Bracteoles 3 or more, often conspicuous.　　　　　　　　　13

10　Leaves ternately divided.　　　　　　　　　　　　　　　　11
　　Leaves pinnately divided.　　　　　　　　　　　　　　　　12

11　A coarse plant, with far-creeping rhizomes; leaflets broad-
　　　lanceolate to ovate.　　　　　　　　　*Aegopodium podagraria*
　　A slender plant with an edible tuber (earthnut); leaves finely
　　　dissected, with narrow segments.　　　　*Conopodium majus*

12　Lower stem-leaves usually 2-pinnate, radical and upper stem-
　　　leaves 1-pinnate; stem hairy; common on chalk.
　　　　　　　　　　　　　　　　　　　　Pimpinella saxifraga
　　All leaves 1-pinnate; stem glabrous; local　　*Pimpinella major*

13　Umbels often simple; bracteoles pinnatifid; mature fruit with
　　　beak 3-7 cms. long.　　　　　　　*Scandix pecten-veneris*
　　Umbels all compound; bracteoles entire; fruit without long
　　　beak.　　　　　　　　　　　　　　　　　　　　　　　14

14　Bracteoles very narrow (linear, subulate, or setaceous).　　15
　　Bracteoles ovate or ovate-lanceolate.　　　　　　　　　　19

15　Leaves ternate, each division 2-pinnate.　　　　　　　　　16
　　Leaves pinnate.　　　　　　　　　　　　　　　　　　　17

16　Bracteoles 0-5, variable in position and length, not markedly
　　　deflexed.　　　　　　　　　　　　　　*Conopodium majus*
　　Bracteoles 3-4, on outer side of partial umbels, up to 1 cm. long,
　　　markedly deflexed.　　　　　　　　　　*Aethusa cynapium*

17　Small (to 30 cm.) annual weed.　　　　　　　*Torilis arvensis*
　　Tall (more than 60 cm.), robust biennials or perennials.　　18

[1] " Bracts " are at the base of the main umbel; " bracteoles " at the base of
the secondary umbels.

18   Leaves 1-pinnate.   *Heracleum sphondylium*
      Leaves 2-3-pinnate.   *Angelica sylvestris*

19   Plant strongly aromatic.   *Myrrhis odorata*
      Plant not aromatic.   20

20   Stem glabrous.   *Anthriscus caucalis*
      Stem hairy below.   *Anthriscus sylvestris*

21   Bracts pinnatifid.   *Daucus carota*
      Bracts entire.   22

22   Plant hairy.   23
      Plant glabrous.   24

23   Leaves 1-3-pinnate, leaflets 1-2 cms. long.   *Torilis japonica*
      Leaves 1-pinnate, leaflets 5-15 cms. long.   *Heracleum sphondylium*

24   Umbel with rays of very irregular length.   *Petroselinum segetum*
      Umbel not very irregular.   25

25   Plant with strong, characteristic smell; stem solid.   *Sison amomum*
      Plant without strong smell; stem hollow.   *Conopodium majus*

LIST OF THE COMMON HEDGEROW AND
ROADSIDE UMBELLIFERS

NAME	SYNONYMS	ENGLISH NAME	MONTHS OF FLOWREING	HABITAT AND DISTRIBUTION
Aegopodium podagraria		Ground Elder, Bishop's Weed, Gout Weed	5—7	Abundant throughout British Isles on disturbed ground.
Aethusa cynapium		Fool's Parsley	7—8	Common arable weed in England, rarer in Scotland.
Angelica sylvestris		Wild Angelica	7—9	Common in damp places throughout British Isles.
Anthriscus caucalis	A. neglecta A. scandicina A. vulgaris Chaerophyllum anthriscus	Bur Chervil	5—6	Local on light, usually sandy soils; commoner near the sea.
Anthriscus sylvestris	Chaerophyllum sylvestre	Hedge Parsley, " Keck "	4—6	The common early-flowering hedgerow umbellifer in England; occurs throughout the British Isles.
Chaerophyllum temulum		Rough Chervil	6—7	Common, but generally less abundant than the preceding, and flowering later.

NAME	SYNONYMS	ENGLISH NAME	MONTHS OF FLOWERING	HABITAT AND DISTRIBUTION
Conium maculatum		Hemlock	6—7	Usually on rather disturbed ground; throughout most of British Isles.
Conopodium majus	C. denudatum, Bunium flexuosum	Earthnut, Pignut	5—6	Fields and woods, throughout British Isles; avoids the chalk and fenland.
Daucus carota		Wild Carrot	6—8	Throughout British Isles, except extreme N. Scotland; very common on chalk.
Foeniculum vulgare	F. officinale	Fennel	7—10	Probably native on English and Welsh coasts; naturalized inland.
Heracleum sphondylium		Cow Parsley, Hogweed, " Keck "	6—9	Common throughout British Isles.
Myrrhis odorata		Sweet Cicely	5—6	Common in N. England and S. Scotland.
Pastinaca sativa	Peucedanum sativum	Wild Parsnip	7—8	Common on chalk in England; rare in Scotland and Ireland.
Petroselinum segetum	Carum segetum	Corn Caraway	7—9	Local in hedgerows, etc.; S. England and S. Wales.

NAME	SYNONYMS	ENGLISH NAME	MONTHS OF FLOWERING	HABITAT AND DISTRIBUTION
Pimpinella major	P. magna	Greater Burnet Saxifrage	6—7	Hedgerows and wood margins; not common, though widely scattered in England and Scotland; S.W. Ireland.
Pimpinella saxifraga		Burnet Saxifrage	7—8	Scattered throughout British Isles; common on chalk grassland.
Scandix pecten-veneris		Shepherd's Needle	4—7	A common arable weed in England; rarer in Wales and Scotland.
Silaum silaus	Silaus flavescens S. pratensis	Pepper Saxifrage	6—8	Meadows and roadsides; locally common in England, absent from N. Scotland and Ireland.
Sison amomum		Stone Parsley	7—9	Hedgerows and roadsides in S. England, the Midlands, and Wales.
Smyrnium olusatrum		Alexanders	4—6	Said to be introduced, but very common in coastal districts of England, and spreading in hedgerows, etc., in many inland areas.

NAME	SYNONYMS	ENGLISH NAME	MONTHS OF FLOWERING	HABITAT AND DISTRIBUTION
Torilis arvensis	Caucalis arvensis Torilis infesta	Spreading Hedge Parsley	7—9	Arable fields; local, chiefly in S. England.
Torilis japonica	T. anthriscus Caucalis anthriscus	Upright Hedge Parsley	7—8	Common in hedgerows throughout British Isles, except N. Scotland; in flower when *Anthriscus* is withered.
Torilis nodosa	Caucalis nodosa	Knotted Hedge Parsley	5—7	Arable fields, roadsides, etc.; locally common in S. England, rare in N. and W. Britain.

KEY TO COMPOSITES WITH YELLOW, "DANDELION-LIKE" FLOWERS[1]

1 Flowering stems without normal foliage leaves (rarely with small reduced ones), with or without small scale-like bracts.　2
　Flowering stems bearing normal foliage leaves.　10

2 Leaves with long stiff white hairs; plant with creeping leafy stolons.　*Hieracium pilosella*
　Leaves not as above; plant without stolons　3

3 Florets more or less equal in length to the involucral bracts.
　　Hypochaeris glabra
　Florets considerably longer than the involucral bracts.　4

4 Flowering stems usually branched, or, if unbranched, with scattered, scale-like bracts.　5
　Flowering stems unbranched, with bracts, if present, only immediately below the flower-heads.　8

5 Florets mixed with linear-lanceolate scales on receptacle.[2]
　　Hypochaeris radicata
　No scales on receptacle.　6

6 Involucral bracts in two distinct rows, the outer much shorter than the inner.　*Crepis capillaris*
　Involucral bracts in several rows.　7

7 Basal leaves usually lobed or pinnatifid, if simple, then linear or lanceolate; flowering stem with numerous small bracts near the heads.　*Leontodon autumnalis*
　Basal leaves usually simple, broader than lanceolate; bracts on flowering stems not numerous near the heads.
　　Hieracium spp.[3]

8 Flowering stems hollow, smooth and shiny, with copious milky juice when broken.　*Taraxacum* spp.[3]
　Flowering stems not as above.　9

[1] i.e. the yellow-flowered members of the sub-family *Cichorioideae* (*Liguliflorae*). Coltsfoot (*Tussilago farfara*), though it is superficially like a dandelion, is omitted, as it has disc and ray florets and does not belong to the *Liguliflorae*.
[2] Pull head to pieces and look carefully with lens (see p. 186).
[3] Many apomictic microspecies of *Hieracium* (hawkweed) and *Taraxacum* (dandelion) have been described (see Clapham, Tutin and Warburg, 1962).

9   Flowering stems usually densely hairy, especially in the upper
     part; fruits all with feathery pappus.   *Leontodon hispidus*
     Flowering stems glabrous, or slightly hairy especially in the
     lower part; central fruits with pappus, outermost fruits with
     only a ring of small scales at the top.   *Leontodon leysseri*

10   Basal leaves grass-like, entire, many times longer than broad.
                                                   Tragopogon pratensis
     Basal leaves not as above.   11

11   Leaves with bristles, each growing on a whitish " blister."
                                                      Picris echioides
     Leaves without " blisters."   12

12   Involucral bracts in two rows, the outer row much shorter than
     the inner.   13
     Involucral bracts in several rows.   19

13   Fruits with no pappus.[4]   *Lapsana communis*
     Fruits with a pappus.   14

14   Inflorescence with many wide-angled branches; involucres
     narrow-cylindrical.   *Mycelis muralis*
     Inflorescence and involucres not as above.   (*Crepis* spp.)   15

15   Leaves, at least the basal ones, pinnatifid.   16
     Leaves entire or toothed.   18

16   Bases of stem-leaves clasping, sagittate; plant glabrous or
     sparingly hairy.   *Crepis capillaris*
     Bases of stem-leaves sometimes clasping, but never sagittate;
     plant hairy.   17

17   Inflorescence sparingly branched, with stout peduncles; fruit
     not beaked.   *Crepis biennis*
     Inflorescence more branched, with slender peduncles; fruit with
     a long beak.   *Crepis vesicaria* subsp. *taraxacifolia*

18   Expanded flower-heads small (usually less than 15 mm. across),
     constricted above directly after flowering; very common.
                                                   Crepis capillaris

[4]This can be determined when the plant is in flower as well as in fruit.

Expanded flower-heads larger (usually more than 20 mm. across), not constricted above; northern marsh plant.

Crepis paludosa

19 Leaves with hooked bristly hairs, especially on the veins beneath. *Picris hieraciodes*

Leaves glabrous, or with simple, branched or glandular hairs. 20

20 Leaves or petioles hairy, usually obviously so, never prickly; milky juice not evident in broken stems. *Hieracium* spp.[3]

Leaves and petioles more or less glabrous (occasionally with sparse hairs), often prickly on the margins and midrib below; broken stems with evident milky juice. 21

21 Involucre narrow-cylindrical, flower-heads in compound panicles. (*Lactuca* spp.) 22

Involucre with broad base, narrowing somewhat upwards; flower-heads in more or less flat-topped groups.

(*Sonchus* spp.) 24

22 Upper stem-leaves linear, clasping, sagittate at base; rather rare, usually near sea. *Lactuca saligna*

Upper stem-leaves not linear, clasping but not sagittate at base. 23

23 Stem-leaves usually placed vertically and orientated north-south ("compass" plant); fruits grey-black, rough.

Lactuca serriola

Stem-leaves placed more or less horizontally; fruits black, smooth; local, usually near sea. *Lactuca virosa*

24 Tall (1 to 2 m.) perennial, with slender, underground stolons; expanded flower-heads 3 to 5 cm. *Sonchus arvensis*[5]

Medium sized (0.5 to 1 m.) annuals or biennials; expanded flower-heads usually 2 to 3 cm. 25

25 Auricles of stem-leaves rounded, appressed; leaves usually distinctly prickly. *Sonchus asper*

Auricles of stem-leaves pointed, ± spreading; leaves not prickly.

Sonchus oleraceus

[5] *Sonchus palustris*, with *pointed* (not *rounded*) auricles to leaves, is a rare marsh plant; *S. arvensis*, which can occur in wet places, is frequently mistaken for it.

LIST OF BRITISH COMPOSITES WITH YELLOW, "DANDELION-LIKE" FLOWERS
(SUB-FAMILY CICHORIOIDEAE OR LIGULIFLORAE)[1]

NAME	ENGLISH NAME	MONTHS OF FLOWERING	HABITAT AND DISTRIBUTION
Arnoseris minima	Lamb's Succory Swine's Succory	6—8	Sandy arable fields, in E. of Gt. Britain; rare.
Crepis biennis	Rough Hawk's-beard	6—7	Fields, waste ground, etc.; local, usually on lowland calcareous soils, in Gt. Britain and Ireland.
Crepis capillaris	Smooth Hawk's-beard	6—9	Common on waste ground, grassland, etc., throughout the British Isles. Extremely variable in size and leaf-shape.
Crepis foetida	Stinking Hawk's-beard	6—8	Roadsides and waste ground; S.E. England, rare.
Crepis mollis	Soft Hawk's-beard	7—8	Damp, shady places in N. Wales, N. England and Scotland; rare.
Crepis paludosa	Marsh Hawk's-beard	6—9	Marshy places; common in N. England and Scotland; Ireland.
Crepis vesicaria subsp. *taraxacifolia*	Beaked Hawk's-beard	5—7	Locally common on waste calcareous ground in S. and C. England and Wales; Ireland; has spread rapidly in last century.
Hieracium pilosella	Mouse-ear Hawkweed	5—9	Common on grassland, heaths, walls, etc., throughout British Isles.

[1] A few rare species are given in the list which are not included in the key, but certain rare introduced species have been omitted from both.

NAME	ENGLISH NAME	MONTHS OF FLOWERING	HABITAT AND DISTRIBUTION
Hieracium spp.	Hawkweeds	—	Many apomictic microspecies in a large variety of habitats throughout British Isles, from sand-dunes and roadsides to the tops of mountains.
Hypochaeris glabra	Smooth Cat's-ear	6—10	Local on non-calcareous sandy soils, particularly fixed dunes; scattered in Gt. Britain; rare in Ireland.
Hypochaeris maculata	Spotted Cat's-ear	6—8	Rare; in chalk grassland and on grassy sea-cliffs, in scattered localities in England and Wales.
Hypochaeris radicata	Cat's-ear	6—9	Common in grassland, on roadsides, etc., throughout British Isles.
Lactuca saligna	Least Lettuce	7—8	Rare, chiefly near the sea; S. and S.E. England only.
Lactuca serriola	Prickly Lettuce	7—9	Waste-ground, dunes, etc.; locally common in S. England and Wales.
Lactuca virosa	Wild Lettuce	7—9	Dunes, waste-ground, etc.; local, scattered throughout Gt. Britain.
Lapsana communis	Nipplewort	6—9	Roadsides and waste places; common throughout British Isles.
Leontodon autumnalis	Autumnal Hawkbit	6—10	Fields, roadsides, etc.; common throughout British Isles and occurring high on mountains (in a dwarfed form).
Leontodon hispidus	Rough Hawkbit	6—9	Grassland, especially on calcareous soils; throughout British Isles, locally abundant

NAME	ENGLISH NAME	MONTHS OF FLOWERING	HABITAT AND DISTRIBUTION
Leontodon leysseri	Hairy Hawkbit	6—9	Dry grassland, especially on calcareous soils and fixed dunes; throughout much of British Isles.
Mycelis muralis (*Lactuca muralis*)	Wall Lettuce	7—9	Walls, rocks, dry woods, usually on calcareous rock; scattered throughout much of British Isles.
Picris echioides	Bristly Ox-tongue	6—10	Roadsides and waste places, especially on heavy, calcareous soils; locally common in S. England, but scattered north to C. Scotland; Ireland.
Picris hieracioides	Hawkweed Ox-tongue	7—10	Roadsides, grassy slopes, etc., especially on calcareous soils; locally common in lowland Britain, extending to S. Scotland; Ireland, rare.
Scorzonera humilis		5—7	Marshy fields; rare, Dorset and Warwickshire only.
Sonchus arvensis	Field Milk-thistle Field Sow-thistle	7—10	Native on sand-dunes, by streams, etc., and as arable weed throughout British Isles, locally very common.
Sonchus asper	Spiny Milk-thistle Spiny Sow-thistle	6—8	Common arable weed throughout British Isles.
Sonchus oleraceus	Milk-thistle Sow-thistle	6—8	Common arable weed throughout British Isles.
Sonchus palustris	Marsh Sow-thistle	7—9	Fens, river-sides; rare, in S.E. England only.

NAME	ENGLISH NAME	MONTHS OF FLOWERING	HABITAT AND DISTRIBUTION
Taraxacum spp.	Dandelion	Mainly early in year	Many apomictic microspecies, found in arable land, on road-sides, in fens and marshes, and on heaths and mountains.
Tragopogon pratensis	Goat's-beard Jack-go-to-bed-at-noon	6—7	Fields, roadsides, etc.; locally common throughout most of British Isles.

BIBLIOGRAPHY

(Dates in brackets indicate reprintings)

ABBOT, C. 1798. *Flora Bedfordiensis*. Bedford.

ARBER, A. 1920. *Water Plants*. Cambridge.

ASHBY, M. 1961 (Ed. 2, 1969). *Introduction to Plant Ecology*. London.

AVEBURY, LORD. 1905. *Notes on the Life-history of British Flowering Plants*. London.

BABINGTON, C. C. 1843. *Manual of British Botany*. London.—1860. *Flora of Cambridgeshire*. London.—1922. *Manual of British Botany*. Ed. A. J. Wilmott. London.

BAKER, H. G. 1947. *Melandrium* in Biological Flora of the British Isles. *J. Ecol.* *35*, 271.

BAUHIN, C. 1623. *Pinax Theatri Botanici*. Basle.

BENTHAM, G. 1858. *Handbook of the British Flora*. London.

BENTHAM, G. and HOOKER, J. D. 1924. ·Ed. 7, rev. by A. B. Rendle, 1947. *Handbook of the British Flora*. London.

BIOLOGICAL FLORA of the British Isles, 1941. Published in *J. Ecol.* and also issued separately.

BONNIER, G. 1917 (1952). *Name this Flower*. Trans. G. S. Boulger. London.

BOSWELL, J. T. I. *See* Syme, J. T. I. Boswell, 1863-92.

BOWLES, E. A. 1934. *A Handbook of Narcissus*. London.

BRIGGS, D. and WALTERS, S. M. 1969. *Plant Variation and Evolution*. London.

BRITTEN, J. and BOULGER, G. S. 1931. *A Biographical Index of Deceased British and Irish Botanists*. London.

BRITTEN, J. and HOLLAND, R. 1878-79. *A Dictionary of English Plant-names*. English Dialect Soc. London.

BUTCHER, R. W. and STRUDWICK, F. E. 1930 (1946). *Further Illustrations of British Plants*. London.

BUTCHER, R. W. 1961. *A New Illustrated British Flora*. London.

CAMDEN, W. 1695. *Britannia*. Ed. E. Gibson. London.

BIBLIOGRAPHY   225

CHAPMAN, V. J. 1938. Studies in Salt-Marsh Ecology. *J. Ecol. 26*, 144.

CHRISTY, M. 1897. *Primula elatior* in Britain. *J. Linn. Soc. (Bot.) 33*, 172.

CLAPHAM, A. R., TUTIN, T. G. and WARBURG, E. F. 1952 (Ed. 2, 1962). *Flora of the British Isles*. Cambridge.—1957-65. *Illustrations to the Flora of the British Isles* (by S. J. Roles). Cambridge.—1959. *Excursion Flora of the British Isles*. Cambridge.

CLARKE, W. A. 1900. *First Records of British Flowering Plants*. London.

CONWAY, V. M. 1954. Stratigraphy of . . . Blanket Peats. *J. Ecol. 42*, 117.

COSTE, H. 1901-6. *Flore . . . de la France*. . . . 3 vols. Paris.

CURTIS, W. 1775-98. *Flora Londinensis*. London.

DANDY, J. E. 1958. *List of British Vascular Plants*. London.

DARLINGTON, C. D. 1956 (Ed. 2, 1963). *Chromosome Botany*. London.

DARWIN, C. 1859. *Origin of Species*. . . . London.—1875. *Insectivorous Plants*. London.

DAVIS, P. H. and HEYWOOD, V. H. 1963. *Principles of Angiosperm Taxonomy*. Edinburgh.

DEERING, G. C. 1738. *Catalogue of Plants . . . Nottingham*. Nottingham.

DILLENIUS, J. J. 1741. *Historia Muscorum*. Oxford.

DONY, J. G. 1953. *Flora of Bedfordshire*. Luton.

DRUCE, G. C. 1886 (Ed. 2, 1927). *The Flora of Oxfordshire*. Oxford.—1897. *The Flora of Berkshire*. Oxford.—1926. *The Flora of Buckinghamshire*. Arbroath.— 1928. *British Plant List*. Arbroath.—1930. *The Flora of Northamptonshire*. Arbroath.—1930 (Ed. 19, 1943). *Hayward's Botanist's Pocket-Book*. London.— 1932. *The Comital Flora of the British Isles*. Arbroath.—1933. Local Floras. *B.E.C. Rep. 1932*, 399.

EDLIN, H. L. 1951. *British Plants and their Uses*. London.

ELLIS, E. A. 1965. *The Broads*. London.

FITCH, W. H. and SMITH, W. G. 1924 (1949). *Illustrations of the British Flora*. London.

FORBES, E. 1846. On the connection between the distribution of the existing Fauna and Flora of the British Isles and the geological changes. . . . *Mem. Geol. Survey U.K. 1*, 336.

GERARD, J. 1597 (Ed. 2 (Ed. T. Johnson), 1633). *The Herball*. London.

GILBERT-CARTER, H. 1936. *British Trees and Shrubs*. Oxford.—1964 (Ed. 3). *Glossary of the British Flora*. Cambridge.

GILMOUR, J. S. L. 1933. The taxonomy of plants intermediate between *Medicago sativa* L. and *Medicago falcata* L. . . . *B.E.C. Rep. 1932*, 393.—1944 (1946). *British Botanists*. London.

GODWIN, H. 1923. Dispersal of Pond Floras. *J. Ecol. 11*, 160.—1944. Age and Origin of the Breckland Heaths of East Anglia. *Nature, 154*, 6.—1949. The Spreading of the British Flora considered in relation to conditions of the Late-Glacial period. *J. Ecol. 37*, 140.—1956. *The History of the British Flora*. Cambridge.

GODWIN, H. and WALTERS, S. M. 1967. The Scientific Importance of Upper Teesdale. *Proc. Bot. Soc. Brit. Is., 6*, 348.

GOOD, R. D'O. 1944. On the distribution of the Primrose in a Southern County. *Naturalist*, No. *809*, p. 41.—1948. *A Geographical Handbook of the Dorset Flora*. Dorchester.

GORDON, H. SETON. 1950. Snow Flora of the Scottish Hills. *Nature, 165*, 132.

GREEN, J. R. 1914. *A History of Botany in the United Kingdom*. London.

GREGOR, J. W. 1934-9. Experimental Taxonomy. *New Phyt. 35*, 323; *37*, 15; *38*, 293.

GUNTHER, R. T. 1922. *Early British Botanists*. . . . Oxford.

GUPPY, H. B. 1893. The River Thames as an Agent in Plant Dispersal. *J. Linn. Soc. (Bot.)*, *29*, 333.

HEGI, G. 1906-66 (Ed. 2 in course of issue). *Illustrierte Flora von Mittel-Europa*. Munich.

HEPBURN, I. 1952. *Flowers of the Coast*. London.

HESLOP-HARRISON, J. 1953. *New Concepts in Flowering-Plant Taxonomy*. London.

HEYWOOD, V. H. 1967. *Plant Taxonomy*. London.

HILL, J. 1760. *Flora Britannica*. London.

HOOKER, J. D. 1870 (Ed. 3, 1884). *The Students' Flora*. London.

HOOKER, W. J. 1830. *British Flora*. London.

HUBBARD, C. E. 1954 (Ed. 2, 1968). *Grasses*. London.

HUDSON, W. 1762. *Flora Anglica*. London.

JACKSON, B. D. 1881. *Guide to the Literature of Botany*. London.—1928 (1949). *A Glossary of Botanic Terms*. . . . London.

JEFFREY, C. 1968. *An Introduction to Plant Taxonomy*. London.

JERMY, C. and TUTIN, T. G. 1968. *British Sedges*. London.

JOHNS, C. A. 1949. *Flowers of the Field*. Ed. R. A. Blakelock. London.

JOHNSON, T. 1629. *Iter . . . in agrum Cantianum*. London.—1632. *Descriptio Itineris . . . in agrum Cantianum*. London.—1634 and 1641. *Mercurius Botanicus*. Parts I and II. London.

JOHNSON, T. 1972. Facsimile edition, with translation by C. E. Raven, entitled *Thomas Johnson—Botanical Journeys in Kent and Hampstead*. Pittsburgh, U.S.A.

KEBLE MARTIN, W. 1965. *The Concise British Flora in Colour*. London.

KEW, H. W. and POWELL, H. E. 1932. *Thomas Johnson, Botanist and Royalist*. London.

KNUTH, P. (1906-9). *Handbook of Flower Pollination*. Trans. J. R. Ainsworth Davis. Oxford.

LANKESTER, E. 1846. *Memorials of John Ray*. . . . London.

LAWRENCE, G. H. M. 1951. *Taxonomy of Vascular Plants*. New York.

LEES, E. 1842 (Ed. 2, 1851). *The Botanical Looker-out*. London.—1843. *Botany of the Malvern Hills*. London.

LEWIS, P. 1958. *British Wild Flowers* (Kew Series). London.

LID, J. 1963. *Norsk og Svensk Flora*. Oslo.

LINDLEY, J. 1829. *Synopsis of the British Flora*. London.

LONG, H. C. 1910. *Common Weeds of the Farm and Garden*. London.—1924. *Plants Poisonous to Live Stock*, Cambridge.

LOUSLEY, J. E. 1939 and 1944. Notes on British *Rumices* I and II. *B.E.C. Rep. 1938*, 118; *1941-2*, 547.—1950. *Wild Flowers of Chalk and Limestone*. London.

LYSAGHT, A. 1959. *Directory of Natural History and other Field Study Societies*. London.

MCLEAN, R. C. and COOK, W. R. I. 1946 (1950). *Practical Field Ecology*. London.

MCCLINTOCK, D. and FITTER, R. S. R. 1956. *Pocket Guide to Wild Flowers*. London.

MAKINS, F. 1939 (1952). *Common Flowers of Britain*. Oxford.

MARSDEN-JONES, E. M. 1930. The genetics of *Geum intermedium*. . . . *J. Genet. 23*, 377.

MARSDEN-JONES, E. M. and TURRILL, W. B. 1930-45. Reports on the Transplant Experiments of the British Ecological Society at Potterne, Wilts. *J. Ecol. 18*, 352; *21*, 268; *23*, 443; *25*, 189; *26*, 359; *33*, 57.

MARTYN, J. 1727. *Methodus Plantarum circa Cantabrigiam nascentium*. London.—1732. *History of Plants growing round Paris*. London.

MATTHEWS, J. R. 1937. Geographical Relationships of the British Flora. *J. Ecol.* *25*, 1.—1955. *Origin and Distribution of the British Flora.* London.

MEIKLE, R. D. 1949. H. C. Watson. *Watsonia, 1,* 3.—1958. *British Trees and Shrubs* (Kew Series). London.

MERRET, C. 1666 (Ed. 2, 1667). *Pinax Rerum Naturalium Britannicarum.* London.

METCALFE, G. 1950. The Ecology of the Cairngorms. Part II, The Mountain Callunetum. *J. Ecol. 38,* 46.

MILLER, P. 1731. *The Gardener's Dictionary.* London.

MOSS, C. E. 1913. *Vegetation of the Peak District.* Cambridge.—1914 and 1920. *The Cambridge British Flora.* Vols. 2 and 3 (all published). Cambridge.

OLIVER, F. W. 1913. *Makers of British Botany.* Cambridge.

PARKINSON, J. 1629. *Paradisi in Sole Paradisus Terrestris.* London.—1640. *Theatrum Botanicum.* London.

PEACE, T. R. and GILMOUR, J. S. L. 1949. The Effect of Picking on the Flowering of Bluebell. *New Phyt. 48,* 115.

PEARSALL, W. H. 1950. *Mountains and Moorlands.* London.

PENNINGTON, W. 1969. *The History of British Vegetation.* London.

PERRING, F. H. (1969). Local Floras currently available for sale. *Proc. Bot. Soc. Brit. Is.* 7, 615.

PERRING, F. H. (Ed.) 1970. The Flora of a Changing Britain. *B.S.B.I. Conference Report 11.* London.

PERRING, F. H. and SAVIDGE, J. P. 1968. Network Research. *Proc. Bot. Soc. Brit. Is.* 7, 553.

PERRING, F. H. and SELL, P. D. 1968. *Critical Supplement to Atlas of the British Flora.* London.

PERRING, F. H. and WALTERS, S. M. 1962. *Atlas of the British Flora.* London.

PIGOTT, C. D. 1954. Species delimitation . . . in British *Thymus. New Phyt. 53,* 470.

PRAEGER, R. L. 1950. *The Natural History of Ireland.* London.

PRIME, C. T. 1960. *Lords and Ladies.* London.

PRIME, C. T. and DEACOCK, R. J. 1935 (Ed. 6, 1970). *Trees and Shrubs. Their identification in Summer and Winter.* Cambridge.—1948. *Shorter British Flora.* London.

PUGSLEY, H. W. 1930. A Revision of the British *Euphrasiae. J. Linn. Soc. (Bot.), 48,* 467.—1948. A Prodromus of the British *Hieracia. J. Linn. Soc. (Bot.), 54,* 1.

RAVEN, C. E. 1947. *English Naturalists from Neckam to Ray.* Cambridge.—1950. *John Ray, Naturalist.* Cambridge.

RAVEN, J. and WALTERS, S. M. 1956. *Mountain Flowers.* London.

RAY, J. 1660. *Catalogus Plantarum circa Cantabrigiam nascentium.* Appendices, 1663 and 1685. Cambridge.—1670 (Ed. 2, 1677). *Catalogus Plantarum Angliae.* London.—1690 (Ed. 2, 1696. Ed. 3 (by Dillenius), 1724). *Synopsis methodica stirpium britannicarum.* London.

REID, E. M. and CHANDLER, M. E. J. 1933. *The London Clay Flora.* London.

RIDDLESDELL, H. J., HEDLEY, R. W. and PRICE, W. R. 1948. *Flora of Gloucestershire.* Arbroath.

RISHBETH, J. 1948. The Flora of Cambridge Walls. *J. Ecol. 36,* 136.

ROSS-CRAIG, S. 1948→. *Drawings of British Plants.* London.

SALISBURY, E. J. 1932. The East Anglian Flora. *Trans. Norf. Norw. Nat. Soc. 12,* 191.—1942. *The Reproductive Capacity of Plants.* London.—1952. *Downs and Dunes.* London.—1961. *Weeds and Aliens.* London.

SCHOTSMAN, H. D. 1949. ... het geslacht *Alisma* in Nederland. *Ned. Kruidk. Arch. 56*, 199.

SIMPSON, N. D. 1960. *A Bibliographical Index of the British Flora*. Bournemouth.

SMITH, J. E. 1790-1814. *English Botany*. Ill. J. Sowerby. 36 vols. London. *See also* Syme, J. T. I. B.

SOWERBY'S *English Botany, see* Smith, J. E. 1790-1814, and Syme, J. T. I. B. 1863-92.

STAMP, L. D. 1969. *Nature Conservation in Britain*. London.

STEP, E. 1940 (1951). *Wayside and Woodland Trees*. Rev. by A. K. and A. B. Jackson. London.

SUMMERHAYES, V. S. 1951. *Wild Orchids of Britain*. London.

SYME, J. T. I. BOSWELL. 1863-92. *English Botany*. Ed. 3. London.

TANSLEY, A. G. 1911. *Types of British Vegetation*. Cambridge.—1939 (1953). *The British Islands and their Vegetation*. Cambridge.—1946 (1949). *Introduction to Plant Ecology*. London.—1948. British Woodlands. *The New Naturalist, A Journal, 1948*, 16.

TANSLEY, A. G. and PRICE EVANS, E. 1946. *Plant Ecology and the School*. London.

TRIMEN, H. and DYER, W. T. T. 1869. *Flora of Middlesex*. London.

TURRILL, W. B. 1948. *British Plant Life*. London.

TUTIN, T. G. et al. (Ed.) 1964, 1968. *Flora Europaea*, Vols. 1 and 2. Cambridge.

VALENTINE, D. H. 1939. A flower problem that readers may help to solve: The Butterbur. *Discovery* (New Series), *14*, 246.—1946. The Butterbur in Yorkshire. *Naturalist*, No. *817*, p. 45.—1947a. The distribution of the sexes in Butterbur. *North Western Naturalist. 1947*, 111.—1947b. Studies in British Primulas. I. Hybridisation between Primrose and Oxlip. ... *New Phyt. 46*, 229.

WARNER, R. 1771. *Plantae Woodfordienses*. London.

WATSON, H. C. 1832. *Outlines of the Geographical Distribution of British Plants*. Edinburgh.—1847-59. *Cybele Britannica*. London.—1873-4. *Topographical Botany*. Thames Ditton.—1883. *Topographical Botany*, Ed. 2. London.

WATT, A. S. 1940. Studies in the Ecology of Breckland. IV. The Grass Heath. *J. Ecol. 28*, 42.

WHITE, J. W. 1912. *The Bristol Flora*. Bristol.

WILLIS, J. C. 1966 (Ed. 7, revised by H. K. Airy Shaw). *A Dictionary of Flowering Plants and Ferns*. Cambridge.

WILSON, J. 1744. *Synopsis of British Plants*. Newcastle-on-Tyne.

WOODELL, S. R. J. 1969. Natural Hybridization in Britain between *Primula vulgaris* Huds. (the Primrose) and *P. elatior* (L.) Hill (the Oxlip). *Watsonia*, 7, 115.

WOODHEAD, N. 1951a. *Lloydia serotina* in Biological Flora of British Isles. *J. Ecol. 39*, 198.—1951b. *Lobelia dortmanna* in Biological Flora of British Isles. *J. Ecol. 39*, 458.

YOUNG, D. P. 1952. Studies in the British *Epipactis*, III and IV. *Watsonia*, 2, 253.

YOUNGMAN, B. J. 1951. Germination of old seeds. *Kew Bull. 1951*, 423.

INDEX-GLOSSARY

All page-references to wild plants are given under their Latin names. Some of the commoner Latin synonyms are also indexed, followed by the name used in this book. English names are indexed under their *first* word, e.g., " great burnet " under " great ", and are followed by their Latin equivalents, under which page-references will be found.

Books are, as a rule, indexed under authors, and not titles. Only place-names of particular botanical interest are indexed. The more important page-references are in bold type; references to illustrations are in italics. Words which have been defined in the text are not, as a rule, defined again. Names of the principal parts of a flower, e.g., petal, sepal, etc., are not defined.

AARON'S ROD, see *Verbascum thapsus*
Abbot, C., 18
Abies alba Mill., 48, 55
achene: a small, dry, indehiscent, single-seeded fruit
Acorus calamus L., *142*, 161
Actaea spicata L., 72, 106
adder's tongue fern, see *Ophioglossum*
Adoxa moschatellina L., 64
Aegopodium podagaria L., 191, 212, 214
Aethusa cynapium L., 212, 214
agave, 30
agg. (aggregate) = sens. lat. (sensu lato); written after the name of a species to indicate that the name, as there used, includes two or more closely allied taxa which have been treated as distinct species (cf. sens. strict.)
Agropyron junceiforme (A. & D. Löve) A. & D. Löve, **168**f, 176; *pungens* (Pers.) Roem. & Schult., 176; *repens* (L) Beauv., 191
Agrostemma githago L., 191
Agrostis, 83, 98
Ajuga reptans L., *39*, 66
Alchemilla, 110; *alpina* L., 117, 119, **121**f, *130*; *arvensis* see *Aphanes arvensis*; *conjuncta* Bab., 122; *faeroensis* Bus., 122f; *glabra* Neyg., 125; *glaucescens* Wallr., 108; *vulgaris* agg., 122
alder, see *Alnus glutinosa*
alder-buckthorn, see *Frangula alnus*
alexanders, see *Smyrnium*
algae, 148, 154, 174
alien plants, 198f
Alisma gramineum C. C. Gmel., 162f; *lanceolatum* With., 162f; *natans* see *Luronium*

natans; *plantago-aquatica* L., 162f
Allium, 184; *ursinum* L., 72f
allo-polyploidy, **45**, 163, 174
all-seed, see *Radiola linoides*
Alnus glutinosa (L.) Gaertn., 54, 76f
alpine butterwort, see *Pinguicula alpina*; club-moss, see *Lycopodium alpinum*; meadow, 113, 125; meadow-grass, see *Poa alpina*; saw-wort, see *Saussurea alpina*; sow-thistle, see *Cicerbita alpina*
Althaea officinalis L., 177; *rosea* (L.) Cav., 177
Ammophila arenaria (L.) Link, 39, *159*, 166, 168f
Anacamptis pyramidalis (L.) L. C. Rich., 104
Anagallis, 90; *arvensis* L., 170, 189
Andromeda polifolia L., 140
Anemone nemorosa L., 59, 60, 62, 67; *pulsatilla* L., 203
Angelica sylvestris L., 125, 212, 214
annual meadow grass, see *Poa annua*; plants, 30, 111, 115, 149, 164, 167, 170f, 194, 195f
Antennaria dioica (L.) Gaertn., 95
Anthriscus neglecta Boiss. & Reut., 213, 214; *sylvestris* (L.) Bernh., 58, 79, 212, 213, 214
Anthyllis vulneraria L., 101
Aphanes arvensis L., 111; *microcarpa* Boiss. & Reut., 111
apomixis: reproduction by seed formed without fertilisation; sometimes extended to cover all asexual reproduction, including vegetative, 33, 45, 56, 65, 110, **122**, 186

Aquilegia vulgaris L., 64
Arabidopsis thaliana (L.) Heynh., 195
Arabis turrita L., 197f
Arber, A., 153, 154, 159
Arbutus unedo L., 10
Arctostaphylos uva-ursi (L.) Spreng., 87
Arenaria, 169
Armeria maritima (Mill.) Willd., 125, 127, 176, *176*, 178
Armoracie rusticana Gaertn., Mey. & Schreb., 196
Arnoseris minima (L.) Schweigg. & Koerte, 221
arrowgrass, see *Triglochin*
arrowhead, see *Sagittaria sagittifolia*
Artemisia norvegica Fr., 124
Arum maculatum L., *46*, 80; *neglectum* (Towns.) Ridl., 80
ash, see *Fraxinus excelsior*
Ashby, M., 210
Asperula cynanchica L., 70f; *odorata* L., 70
Asplenium, 197; *viride* L., 125
Aster tripolium L., 175; var. *discoideus* Rehb., 175
Astragalus danicus Tetz., 101f, 187; *glycyphyllos* L., 187f
Atlantic distribution: a distribution in Western Europe, not extending far into Central or Eastern Europe, 55, 62, 87, 95, 168, 169, 180; period, 53, **54**, 56, 129
Atriplex glabriuscula Edmonst., 167; *hastata* L., 167; *laciniata* L., 167; *patula* L., 167
Atropa belladonna L., *190*, 196
autecology, 202
auto-polyploidy, 45
autumn crocus, see *Colchicum autumnale*
autumnal hawkbit, see *Leontodon autumnalis*

Avebury, Lord, 210
Avena pratensis see *Helictotrichon pratense*
Avon Gorge, 98, 198
awlwort, see *Subularia aquatica*
Azolla, 155
BABINGTON, C. C., 13, 21, 22, 209
back-cross: fertilisation between the progeny of a cross and one of the parents
Baker, H. G., 190
Baldellia ranunculoides (L.) Parl., *143*, 162
Ballota nigra L., 80
balsam, see *Impatiens*
bamboo, 30
baneberry, see *Actaea spicata*
Banks, J., 19
Barbarea vulgaris R. Br., 196
barren strawberry, see *Potentilla sterilis*
Bartsia alpina L., 14
Bauhin, C., 16
beaked hawks-beard, see *Crepis vesicaria* subsp. *taraxacifolia*
bearberry, see *Arctostaphylos uva-ursi*
beech, see *Fagus sylvatica*
bee orchid, see *Ophrys apifera*
bell-heather, see *Erica cinerea*
Bellis perennis L., 40, 44
Ben Lawers, 107, 111, 121
bent, see *Agrostis*
Bentham, G., 20, 22, 209
Berry Head, 102, 105
Beta maritima L., 168
betony, see *Stachys officinalis*
Betula, 52, 53, 76, 84, 114, 129; *tenta* L., 75; *pendula* Roth, 74; *pubescens* Ehrh., 74
biennial, *30*, 167, 195f
bilberry, see *Vaccinium myrtillus*
bindweed, see *Convolvulus arvensis*
Biological Flora of the British Isles, 27, 202
birch, see *Betula*
birchwood, *74*, 114, 129
bird's-eye primrose, see *Primula farinosa*
bird's-foot trefoil, see *Lotus corniculatus*
bird's nest orchid, see *Neottia nidus-avis*
bishop's weed, see *Aegopodium podagraria*
biting stonecrop, see *Sedum acre*

black horehound, see *Ballota*
Blackstone, W., 17
bladder campion. see *Silene vulgaris*
bladderwort, see *Utricularia*
Blakeney Point, 168, 169, 172, 173
blanket bog, 37, 82, 129, 132, *133f*
Blechnum spicant (L.) Roth, 93
bloody cranesbill, see *Geranium sanguineum*
bluebell, see *Endymion*
blue moor-grass, see *Sesleria*
bog, 37, 82, 129, 131, 132ff, 142, 147, 152; asphodel, see *Narthecium ossifragum*; whortleberry, see *Vaccinium uliginosum*
bombed site plants, 194
Bonnier, G., 209
Boreal period, 53ff, 76, 110, 129
Boswell, J. T. I., see *Syme, J. T. I. B.*
Botanical Society of the British Isles (B.S.B.I.), 4f, 25, 201, 202, 204
Botrychium lunaria (L.) Sw., 93f
Boulger, G. S., 210
Bowles, E. A., 185
Bowles, G., 10
box, see *Buxus sempervirens*
Box Hill, 96
bracken, see *Pteridium aquilinum*
Brachypodium pinnatum (L.) Beauv., 99; *sylvaticum* (Huds.) Beauv., 99
bract: a leaflike organ in an inflorescence
bracteole: a small bract
bramble, see *Rubus fruticosus*
Brassica oleracea L., 158, **178f**
Braunton Burrows, 168
Brean Down, 102
Breckland, 53, 76, **84f**, 90, 91, 94, 99
Brewer, S., 16
Briggs, D., 151, 210
bristly ox-tongue, see *Picris echioides*
Britten, J., 210
Briza media L., 98
Broads, 76, 142, 143, 149, **152**, 160
Bromus erectus see *Zerna erecta*
brookweed, see *Samolus*
broom, see *Sarothamnus*
broomrape, see *Orobanche*
Brown, Littleton, 16

buckthorns, see *Rhamnus cathartica* and *Frangula alnus*
Buddle, A., 15
Buddleia davidii Franch., 194
bugle, see *Ajuga reptans*
bulb: an underground storage organ containing next year's bud and a stem surrounded by a number of fleshy leaf-bases or scale leaves
bulbil: a small bulb or tuber on the aerial part of a plant, e.g., in a leaf axil or in an inflorescence
bulrush, see *Typha*, and *Schoenoplectus lacustris*
Bunium bulbocastanum L., 105
Buplerum opacum (Ces.) Lange, 105; *rotundifolium* L., 105
bur chervil, see *Anthriscus caucalis*
burnet rose, see *Rosa spinosissima*; saxifrage, see *Pimpinella saxifraga*
bur-reed, see *Sparganium*
Burren, 134
Butcher, R. W., 66, 209
Butomus umbellatus L., 8, 159
butterbur, see *Petasites hybridus*
buttercup, see *Ranunculus*
butterwort, see *Pinguicula*
Buxus sempervirens L., 109
CABBAGE, see *Brassica oleracea*
Cactaceae, 191
Cairngorms, 87, 114f, 116f, 121
Cakile maritima Scop., 166f
calcicole: usually growing on, or confined to, limy soils
calcifuge: not usually growing on limy soils
Calluna vulgaris (L.) Hull, 39, 67, 74, 81, **82**, 84, 85, **86f**, 95, 98, 124
Caltha palustris L., 58, 183
Calystegia sepium (L.) Roem. & Schult., 76, 78; *soldanella* (L.) R. Br., *170*; *sylvestris* (Willd.) Roem. & Schult., 78
Cambridge, 7, 10, 12ff, 15, 101, 143, 175, 183, 197; Professors of Botany, 16, 22
Camden's "Britannia," 14
Campanula latifolia L., 64; *rotundifolia* L., 89, 125; *trachelium* L., 64
Canadian waterweed, see *Elodea canadensis*

Capsella bursa-pastoris L., 30, 191, 194, 195
Carboniferous, 47f; limestone, 71, 97
Cardamine hirsuta L., 195; *pratensis* L., 34, 183
Cardaria draba (L.) Desv., 196
Carex, **137**ff; *acutiformis* Ehrh., 139; *arenaria* L., 84, 139, 170; *bigelowii* Torr., 118; *binervis* Sm., 139; *caryophyllea* Latour., 99, 139; *disticha* Huds., 138; *elata* All., 138; *ericetorum* Poll., 99, 129; *extensa* Good., 139; *flacca* Schreb., 99, 106, 139; *glauca* see *C. flacca*; *hirta* L., 138; *nigra* (L.) Reichard, 139; *otrubae* Podp., 139; *pendula* L., 19, 66; *praecox* see *C. caryophyllea*; *rigida* see *C. bigelowii*; *riparia* Curt., 139; *rostrata* Stokes, 52; *sylvatica* Huds., 139; *vulpina* see *C. otrubae*
Carpinus betulus L., 67
carr, 76f, 143
carrot, see *Daucus carota*
Carum bulbocastanum see *Bunium bulbocastanum*; *segetum* see *Petroselinum segetum*
cat-mint, see *Nepeta cataria*
cat's-ear, see *Hypochaeris*
Caucalis spp., see *Torilis*
Centaurea cyanus L., 53, 191
Centaurium erythraea Rain, 31, **71**, 171; *littorale* (Turner) Gilmour, 171
centaury, see *Centaurium*
Centunculus minimus L., 89f
Cephalanthera damasonium (Mill.) Druce, 69
Cerastium alpinum L., 126; *articum* Lange, 126; *cerastioides* L.) Britton, 126; *semidecandrum* L., 170; *vulgatum* L., subsp. *fontanum* Baumg., 126
Chaerophyllum temulum L., 79, 211, 214
chaffweed, see *Centunculus minimus*
chalk, 68, 96ff
Chamaenerion angustifolium (L.) Scop., 194, 195
Chandler, M. E. J., 48
Chapman, V. J., 173
charlock, see *Sinapis arvensis*
Cheddar Gorge, 97, 98, 108, 109, 111

Cheddar pink, see *Dianthus gratianopolitanus*
Cheiranthus cheiri L., 158, 198
Chelsea Physic Garden, 15, 18
Chenopodium, 167, 176
Cheyne, Lord, 15
chickweed, see *Stellaria*; wintergreen, see *Trientalis europaea*
chlorophyll: the green colouring matter in plants, 193
Christy, M., 61
chromosome, **42**f, 162, 163, 184, 185, 204
Chrysanthemum leucanthemum L., 125; *segetum* L., 191, 193
Cicendia filiformis (L.) Delarbre, 90
Cicerbita alpina (L.) Wallr., 125
Circaea lutetiana L., 66
Cirsium anglicum see *C. dissectum*; *dissectum* (L.) Hill, 73; *eriophorum* (L.) Scop., subsp. *britannicum* Petrak, 187; *heterophyllum* (L.) Hill, 73; *palustre* (L.) Scop., 73; *pratense* see *C. dissectum*
Cladium mariscus (L.) Pohl, 67, **137**f, 141, 142, 144, 145
clary, wild, see *Salvia horminoides*
Clapham, A. R., xiii, xiv, 207-9
Clarke, W. A., 11, 14
cleistogamous flowers, **34**, 91, 195
Clematis vitalba L., 78f
climate, 36f; changes in British, **48**ff, 95, 104, 142; mountain, 114
climax, biotic, **40**, 182; climatic, **40**, 59; edaphic, 40
climbing plants, 76, 77f
Clinopodium vulgare L., 80
clone: a group of individuals derived vegetatively from a single individual
closed community (habitat): a community in which the plants grow close together and competition between them is heavy, 129
clover, see *Trifolium*
club-moss, see *Lycopodium*
Cochlearia see *Armoracia*
cock's-foot grass, see *Dactylis*
colchicine, 45, 185
Colchicum autumnale L., 184f

collecting, 3f, 6, 205f
coltsfoot, see *Tussilago farfara*
columbine, see *Aquilegia*
comfrey, see *symphytum*
common butterwort, see *Pinguicula vulgaris*; calamint, see *Calamintha ascendens*; lady's mantle, see *Alchemilla vulgaris*; gorse, see *Ulex europaeus*; mallow, see *Malva sylvestris*; mullein, see *Verbascum thapsus*; reed, see *Phragmites communis*; rock-rose, see *Helianthemum chamaecistus*; teasel, see *Dipsacus fullonum* subsp. *sylvestris*
community, plant, 26, 36, 38ff, 129
Compositae, 35, 122, 179, 187; yellow, 3, 71, 170, 186, **218**ff
Conium maculatum L., 211, 215
Conopodium majus (Gouan) Lor. ; Barr., 212, 215
Continental distribution: a distribution mainly in Central and Eastern Europe, 55, 85, 100, 101f
Convallaria majalis L., 72, 102
Convolvulus arvensis L., 182, 191; *sepium* see *Calystegia sepium*; *soldanella* see *Calystegia soldanella*
Cook, C. D. K., 151
Cook, W. R. I., 99, 210
Corallorhiza trifida Chatel., 74
coral-root orchid, see *Corallorhiza trifida*
cord-grass, see *Spartina*
corm: a swollen, bulb-like underground stem
corn caraway, see *Petroselinum segetum*; cockle, see *Agrostemma githago*; gromwell, see *Lithospermum arvense*; marigold, see *Chrysanthemum segetum*
cornflower, see *Centaurea cyanus*
Cornish heath, see *Erica vagans* L.
Coronopus didymus (L.) Sm., 195; *squamatus* (Forsk.) Aschers., 195
Corylus avellana L., 33, 34, 53, 59
Cosmos, 32
Coste, H., 201
cotton-grass, see *Eriophorum*
Cotyledon umbilicus see *Umbilicus rupestris*

couch grass, see *Agropyron repens*
Council for Nature, 5
Council for the Promotion of Field Studies, 106
cow parsley, see *Heracleum*
cowberry, see *Vaccinium vitisidaea*
cowslip, see *Primula veris*
Crabbe, G., 172
Crambe maritima, L., 167
cranberry, see *Oxycossus*
Crataegus, 29
creeping buttercup, see *Ranunculus repens*; willow, see *Salix repens*; yellowcress, see *Roripla sylvestris*
Crepis, 3; *biennis* L., 219, 221; *capillaris* (L.) Wallr., 218, 219, 221; *foetida* L., 221; *mollis* (Jacq.) Aschers., 221; *paludosa* (L.) Moench, 220, 221; *vesicaria* L., subsp. *taraxacifolia* (Thuill.) Thell., 219, 221
Cretaceous, 48
Crithmum maritimum L., 179
"critical" groups, 150, 200f
Crocus nudiflorus Sm., 184; *sativus* L., 184
Cromer forest beds, 48
cross: a loose term applied to fertilisation between two different kinds of plant, usually between different species
Cross, W. R., 143
cross-leaved heath, see *Erica tetralix*
crowberry, see *Empetrum*
crown imperial, see *Fritillaria imperialis*
Cruciferae, 158, 194ff
cuckoo-flower, see *Cardamine pratensis*
cuckoo-pint, see *Arum maculatum*
curled dock, see *Rumex crispus*
Curtis, W., 18f, 167
Cuscuta epilinum Weihe, 193; *epithymum* (L.) Murr., 193; *europaea* L., 193
Cymbalaria muralis Baumg., 174, 197
Cyperaceae, 137ff
Cypripedium calceolus L., 10
cytology, 25f, 42f, 204
Dactylis glomerata, L., 45
daffodil, see *Narcissus*
daisy, see *Bellis perennis*
Dale, S., 16

dandelion, see *Taraxacum*
Dandy, J. E., 209
Daphne laureola L., 70, *mezereum* L., 70
dark mullein, see *Verbascum nigrum*
Darlington, C. D., 210
darnel, see *Lolium temulentum*
Darwin, C., and Darwinism, 20, 26, 41f, 47, 77, 128, 135, 153
Daucus carota L., 105, 143, 179, 213, 215; *gingidium* L., 179
Davis, P. H., 201
de L'Obel, M., see L'Obel
Deacock, R. J., 209
deadly nightshade, see *Atropa*
deciduous: dropping off, used especially for losing leaves in the autumn
deer-grass, see *Trichophorum caespitosum*
Deering, G., 17
Deschampsia caespitosa (L.) Beauv., 124, 146; *flexuosa* (L.) Trin., 89
devil's-bit scabious, see *Succisa pratensis*
Dianthus caesius see *D. gratianopolitanus*; *deltoides* L., 78, 108; *gratianopolitanus* Vill., 108, 109
Diapensia lapponica L., 124
Dillenius, J. J., 15f
dimorphous: having two forms
dioecious plants: plants having the sexes on different individuals; *Empetrum*, 120; *Petasites*, 161; *Salix*, 126; *Stratiotes*, 160; *Trinia*, 105
diploid, 42f, 45, 101, 163, 184
Diplotaxis muralis (L.) DC., 195
Dipsacus, 170; *fullonum* L., 187; *pilosus* L., 187
dispersal of seeds and fruits, 35, 66, 153f, 164, 194, 198
distribution, different for male and female, 160f; groups in British flora, 55; water-plants, 153f
dock, see *Rumex*
dodder, see *Cuscuta*
dog's mercury, see *Mercurialis perennis*
dog violet, see *Viola*
Dony, J. G., 22
Doody, S., 15

Dorset heath, see *Erica ciliaris*
Draba verna see *Erophila verna*
dropwort, see *Oenanthe*
Drosera, 136, 146; *anglica* Huds., 31, 110, 135; *intermedia* Drev. & Heyne, 31, 135; *rotundifolia* L., 31, 135
Druce, G. C., 24f, 208, 209
Dryas octopetala L., 52, 120, 127, 128
Dryopteris thelypteris see *Thelypteris palustris*
duckweed, see *Lemna*
dune-slacks, 145, 171
durmast oak, see *Quercus petraea*
dwarf cudweed, see *Gnaphalium supinum*; willow, see *Salix herbacea*
Dyer, W. T. T., 15, 22
EARTHNUT, see *Conopodium majus*
eco-cline, 44
Ecological Society, British, 26, 27, 202
ecology, 25f, 28, 36ff, 202, 204
ecotype, 26, 44, 67, 98, 100, 103, 125, 127, 173
Edlin, H. L., 210
Edwards, S., 19
Elatine, 152
Eleocharis acicularis (L.) Roem. & Schult., 158; *palustris* (L.) Roem & Schult., 139
Ellis, E. A., 152
elm, see *Ulmus*
Elodea canadensis Michx., 149, 154, 161, 162, 164f
Elymus arenarius L., 168f
Empetrum, 52, 124, 128, 129, 134; *hermaphroditum* (Lange) Hagerup, 119f; *nigrum*, L., 95, 119f
enchanter's nightshade, see *Circaea lutetiana*
endemic pants: plants confined as natives to a specified area, 55f, 110, 187
Endymion nonscriptus (L.) Garcke, 7, 27, 43, 47, 59, 62f, 180
English Botany (Sowerby), 19, 22
Eocene, 48
Epilobium adenocaulon Hausskn., 66; *adnatum* Gris., 66; *lamyi* F. Schultz, 66; *montanum* L., 66; *tetragonum* see *E. adnatum*; *parviflorum* Schreb., 66; *obscurum* Schreb., 66

Epipactis, 69; *palustris* (L.) Crantz, 158, 171
Equisetum, 94; *palustre* L., *146*
Eranthis hiemalis (L.) Salisb., 59, *66*
Erica carnea L., 87; *ciliaris* L., 87; *cinerea* L., 85f, *mackayana* Bab., 87; *tetralix* L., 85f; *vagans* L., 86f
Eriophorum, 49, 131, 134, 139; *angustifolium* Honck., 138; *gracile* Roth, 138; *latifolium* Hoppe, 138; *vaginatum* L., 82, *83*, 138
Erodium maritimum (L.) Ait., 180f
Erophila verna (L.) Chevall., 195
Erynigium maritimum L., 39, 170
Euphorbia amygdaloides L., 63; *paralias* L., 170
Euphrasia, 3, 201
evening primrose, see *Oenothera*
evolution, 20, 26, **41**f, 46, 47, 204
eyebright, see *Euphrasia*
Fagus sylvatica L., 54, **68**ff, 96f
felwort, see *Gentianella amarella*
fen, conservation, 147, 203; drainage, 136, 141, 247; floristic richness, 144; meaning of term, 131; 140f; orchid, see *Liparis loselii*; violet, see *Viola stagnina*
Fenland, 67, 136, **141**f, 152, 159, 160
fennel, see *Foeniculum vulgare*
ferns, 93f, 197
fertilisation: fusion of a male and female gamete, 29
Festuca ovina, L., 83, 93, 106; *rubra* L., 176; *rubra* L., var. *arenaria* (Osb.) Fr., 170
Ficaria verna see *Ranunculus ficaria*
field layer, 58f, 67, 74; milk-thistle, see *Sonchus arvensis*; sow-thistle, see *Sonchus arvensis*
Filipendula ulmaria (L.) Max., 52, 177, 183
fir club-moss, see *Lycopodium selago*
Fitch, W. H., 209
Fitter, R. S. R., 209
flag, see *Iris pseudacorus*
flax, see *Linum usitatissimum*

flea-bane, see *Pulicaria dysenterica*
flora: the kinds of plant occurring in a specified area
flora, British compared with f. of other countries, 21, 22, 27, 54f; distribution groups, 55
Flora: an enumeration, with or without further information, of the plants of a specified area
Flora, British, the first, 10f, 14
Flora Europaea, 201
Flora of the British Isles (Clapham, Tutin & Warburg), xiii, xiv, 22, 69, 110, 201, 207, 209
Floras, local, 3, 4, 9, 12, **18**, 20ff, 208
flowering, conditions of, 32; plants, origin of, 47f, rush, see *Butomus umbellatus*
Foeniculum vulgare Mill., 211, 215
fool's parsley, see *Aethusa cynapium*
Forbes, E., 128
forest, clearance by man, 54; deciduous, 37, 129; evergreen tropical, 27
forget-me-not, see *Myosotis*
Fowler, W., 143
Fragaria chiloensis Duchesne, 71; *vesca* L., 70, 71; *virginiana* Duchesne, 71
fragrant orchid, see *Cymnadenia conopsea*
Frangula alnus Mill., 76, 143
Fraxinus excelsior L., 39, 40, **71**ff, 76, 106
fringed water-lily, see *Nymphoides peltatum*
Fritillaria imperialis L., 183; *meleagris* L., 183
fritillary, see *Fritillaria meleagris*
frog-bit, see *Hydrocharis*
fuller's teasel, see *Dipsacus fullonum*
Gagea lutea (L.) Ker-Gawl., 63
Galeobdolon luteum Huds., 8, *50*, 73
Galium anglicum Huds., 25; *aparine* L., 35; *boreale* L., 14; *verum* L., 170
gamete: a male or female sex cell
garden escape, 194, 196

garlic, wild, see *Allium ursinum*
Gaultheria procumbens L., 75
gene, 33, **42**f
genetics, 25f, **42**f, 204
Genista, 88
Gentianella amarella (L.) H. Sm., *164*, 171
Gentiana nivalis L., 115
Gentianaceae, 90
genus (pl. genera), 40f
Geranium pratense L., 72, 108, *115*; *sanguineum* L., 72; *sylvaticum* L., 108
Gerard, J., **8**, 14
Geum rivale L., 65; *urbanum* L., 65
Gilbert-Carter, H., 209, 210
Gilmour, J. S. I., 61, 210
glabrous: without hairs
Gladiolus, 184
glasswort, see *Salicornia*
Glaucium flavum Crantz, 172
Glechoma hederacea L., 63, 72
globe flower, see *Trollius europaeus*
Glyceria declinata Bréb., 152; *maritima* see *Puccinellia maritima*
Glycyrrhiza glabra L., 188
Gnaphalium supinum L., 118, 119, 123
goatsbeard, see *Tragopogon pratensis*
Godwin, H., 52, 56, 84, 128, 153, 210
golden rod, see *Solidago*
goldilocks, see *Ranunculus auricomus*
Good, R. D'O., 60, 121
Goodyer, J., 9ff
Goodyera repens (L.) R.Br., 9, 74
goosefoot, see *Chenopodium*
goose-grass, see *Galium aparine*
Gordon, H. S., 119
gorse, see *Ulex*
gout weed, see *Aegopodium*
Gramineae, **98**f, 116, 134, 137, 158
grass of Parnassus, see *Parnassia palustris*
grasses, see *Gramineae*
grass-wrack, see *Zostera*
grazing, cattle, 144, 176, 182; deer, 116, 125; rabbit, 84, 94, 96, 99; sheep, 83, **96**f, 106f, 125, 176, 182
great burnet, see *Sanguisorba officinalis*; earth-nut, see *Bunium bulbocastanum*
Great Orme, 97

greater burnet saxifrage, see *Pimpinella major*

green hellebore, see *Helleborus viridis*; spleenwort, see *Asplenium viride*

Green, J. R., 210

Gregor, J. W., 26, 176

Gronovius, J. F., 75

ground elder, see *Aegopodium*

ground ivy, see *Glechoma*

groundsel, see *Senecio vulgaris*

guelder rose, see *Viburnum opulus*

Gunther, R. T., 10, 11, 210

Guppy, H. B., 153

Gymnadenia conopsea (L.) R. Br., 104

gynoecium: the female part of a flower consisting of one or more ovaries with their styles and stigmas

Habenaria conopsea see *Gymnadenia conopsea*

hairy bitter-cress, see *Cardamine hirsuta*; hawkbit, see *Leontodon leysseri*

Halimione portulacoides (L.) Aell., 176f, 178

halophytes: plants adapted to saline conditions, 173, 178

hard fern, see *Blechnum spicant*

harebell, see *Campanula rotundifolia*

hawkbit, see *Leontodon*

hawk's-beard, see *Crepis*

hawkweed, see *Hieracium*; ox-tongue, see *Picris hieracioides*

hawthorn, see *Crataegus*

hazel, see *Corylus avellana*

heath (habitat), 82ff; (plant), see *Ericaceae* and *Erica*; false-brome, see *Brachypodium pinnatum*

heather, see *Calluna vulgaris*; moor, 82f

Hedera helix L., 69

hedge mustard, see *Sisymbrium officinale*; parsley, see *Anthriscus sylvestris*; woundwort, see *Stachys sylvatica*

Hegi, G., 201

Helianthemum apenninum (L.) Mill., 102; *canum* (L.) Baumg., 102, 109, 128; *chamaecistus* Mill., 102, 110; *guttatum* L., subsp. *breweri* (Planch.) Hook, 31; *nummularium* see *H. chamaecistus*; *polifolium* see *H. apenninum*; *vulgare* see *H. chamaecistus*

Helictotrichon pratense (L.) Pilger, 98, 110

Helleborus foetidus L., *34*, 63; *niger* L., 63; *viridis* L., 63

hemlock, see *Conium maculatum*; water dropwort, see *Oenanthe crocata*

Henslow, G., 159

Hepburn, I., 168, 210

Heracleum sphondylium L., 79, 212, 213, 215

herb, Christopher, see *Actaea spicata*; Paris, see *Paris*

herbarium: a collection of dried plants mounted on paper, 6, *175*, **205ff**

Heslop-Harrison, J., 210

heterophylly: the production of two or more types of leaf on one plant, 151

Heywood, V. H., 210

Hieracium, 3, 56, 79, **186**, 201, 220, 222; *pilosella* L., 218, 221

Hill, J., 17

Himalayan balsam, see *Impatiens glandulifera*

Himantoglossum hircinum (L.) Sprengel, 104

Hippocrepis comosa L., 101

Hippothaë rhamnoides L., 171

hoary pepperwort, see *Cardaria draba*; rock-rose, see *Helianthemum canum*

hogweed, see *Heracleum*

Holland, R., 210

hollyhock, see *Althaea rosea*

honewort, see *Trinia glauca*

Honkenya peploides (L.) Ehrh., 169, 172

Hooker, J. D., 20, 22, 209

Hooker, W. J., 21

Hooper, M. D., 77

hop, see *Humulus lupulus*

hornbeam, see *Carpinus betulus*

horned sea poppy, see *Glaucium flavum*

horse radish, see *Armoracia*

horseshoe vetch, see *Hippocrepis comosa*

horsetail, see *Equisetum*

Hottonia palustris L., 159

How, W., 10f

Hubbard, C. E., 99, 175, 209

Hudson, W., 14, 16, 17f, 21

Humulus lupulus L., 76, 193

Hunt Institute of Botanical Documentation, 9

Huxley, J. S., 210

hybridization, *Alisma*, 163;

Centaurium, 171; *Drosera* 135; *Epilobium*, 66; *Fragaria*, 71; *Geum*, 64f; *Medicago*, 61; *Pinguicula*, 137; *Primula*, 61; *Silene*, 190; *Viola*, 91f

Hydrocharis morsus-rance L., 159, 160

hydrosere, 141f, 144, 151

Hypericum hirsutum L., 68; *montanum* L., 72; *perforatum* L., 68; *pulchrum* L., 68

hypha: one of the thread-like bodies which, interwoven together, constitute the "tissue" of a fungus

Hypochaeris glabra L., 170, 218, 222; *radicata* L., 186, 218, 222

ICE AGE, 48ff, 111, 122, 130, 142, 152, 180

Illecebrum verticillatum L., 90

Impatiens, 35; *capensis* Meerburgh, 164; *glandulifera* Royle, 163f; *noli-tangere* L., 164; *parviflora* DC., 164

indehiscent fruit: one not opening to release its seeds

inflorescence: the portion of a plant bearing the flowers

Ingleborough, 14, 106, 108, 134

insectivorous plants, 135f

inter-glacial period, 48, 54

introduced plants, 110f, 154, 155, 163f, 190, 191, 194, 197ff

Inula dysenterica see *Pulicaria dysentarica*

involucre: a calyx-like collection of bracts, e.g., below the flower-head in composites

Ireland, absences from flora, 55, **102**, 175, 188; joined to Britain, 52

Iris, 184; *pseudacorus* L., 7, 148, 161, *162*

Irish butterwort, see *Pinguicula grandiflora*

Isoetes, 156f, 158; *echinospora* Durieu, 157; *hystrix* Durieu, 157; *lacustris* L., 157

ivy, see *Hedera helix*

ivy-leaved bellflower, see *Wahlenbergia hederacea*; crowfoot, see *Ranunculus hederaceus*; duckweed, see *Lemna trisulca*; toadflax, see *Cymbalaria muralis*

JACKSON, B. D., 210

Jacob's ladder, see *Polemonium coeruleum*
Jeffrey, C., 210
Jermy, C., 209
Johns, C. A., 2, 209
Johnson, T., 8ff, 14, 23, 98
Jones, J. P., 22
Juncus, 146; *acutus* L., 171; *trifidus* L., 115
Jurassic, 48
KEBLE MARTIN, W., 209
Keck, see *Anthriscus sylvestris* and *Heracleum sphondylium*
Kew Gardens, 7, 21, 22, 62
Kew, H. W., 8, 10
kingcup, see *Caltha palustris*
Kingston, J. F., 22
knotgrass, see *Polygonum*
knotted hedge-parsley, see *Torilis nodosa*
Knuth, P. 157
Kobresia simpliciuscula (Wahlenb.) Mackenzie, 127
Koeleria gracilis Pers., 98
Koenigia islandica L., 124
Labiatae, 67, **79**
Labrador tea, see *Ledum*
Lactuca muralis see *Mycelis muralis*; *saligna* L., 220, 222; *serriola* L., 220, 222; *virosa* L., 220, 222
ladies' bedstraw, see *Galium verum*
lady's fingers, see *Anthyllis vulneraria*; mantle, see *Alchemilla*; slipper orchid, see *Cypripedium calceolus*; tresses orchid, see *Spiranthes*
lakes, upland, 149, 151, 152, 156, 158
lamb's succory, see *Arnoseris minima*
Lamium galebdolon see *Galeobdolon luteum*; *purpureum* L., 79
lanceolate (leaf): narrow, tapering at each end
Lankester, E., 13
Lapsana communis L., 219, 222
large bindweed, see *Calystegia sepium* and *C. sylvestris*
Late Glacial period, 51ff, 102, 109, 110, 128ff, 145
Lathyrus japonicus Willd., 166, 172; *palustris* L., 143f
Latimer, H., 7
"laurel", 156
Lavatera arborea L., 177, 179
Lawrence, G. H. M., 210
least lettuce, see *Lactuca saligna*

Ledum palustre, L., 140
Lees, E., 23
Lemna, 149, 150; *gibba* L., 155; *minor* L., 153, 154f; *polyrrhiza* L., 155, 160; *trisulca* L., 155
Leontodon, 3; *autumnalis* L., 186, 218, 222; *hirtus* see *L. leysseri*; *hispidus* L., 219, 222; *leysseri* (Wallr.) Beck, 219, 223
Lepidium draba see *Cardaria draba*
lesser broomrape, see *Orobanche minor*; burnet-saxifrage, see *Pimpinella saxifraga*; calamint, see *Calamintha nepeta*; celandine, see *Ranunculus ficaria*; spearwort, see *Ranunculus flammula*
lettuce, wild, see *Lactuca tirosa*
Lewis, P., 209
Lhuyd, E., 16, 123
liane: a vigorous woody climber in a tropical forest
lichens, 119, 197
Lid, J., 201
life-form, 35
Ligusticum scoticum L., 179
lily-of-the-valley, see *Convallaria majalis*
lime, see *Tilia*
limestone, geological nature, 97; grassland, 106ff; pavement, 106, 127, 130
Limnanthemum peltatum see *Nymphoides peltatum*
Limonium, 174, 176; *bellidifolium* (Gouan) Dum., 175; *binervosum* (G.E.Sm.) C. E. Salmon, 180; *humile* Mill., 175; *paradoxum* Pugsl., 180; *recurvum* C. E. Salmon, 180; *transwallianum* (Pugsl.) Pugsl., 180; *vulgare* Mill., 165, 175, 178, 180
Limosella, 149
Linaria cymbalaria see *Cymbalaria muralis*; *vulgaris* Mill., 179, 197
Lindley, J., 21
ling, see *Calluna vulgaris*
Linnaea borealis, L., 75
Linnaeus, C., 17ff, 41, 64, 75, 91, 194, 198
Linnean Society, 19; system, 15, 17f, 20
Linum anglicum Mill., 13; *usitatissimum* L., 193
Liparis loeselii (L.) L. C.

Rich., 144f
liquorice, see *Glycyrrhiza*
liquorice-vetch, see *Astralagus glycyphyllos*
Listera ovata (L.) R.Br., 8
Lithospermum arvense L., 31
Littorella uniflora (L.) Aschers., 156, 158
lizard orchid, see *Himantoglossum hircinum*
Lizard Peninsula, 21, 86, 157
Lloydia serotina (L.) Rchb., 123
L'Obel, M. de, 7f, 14, 63
Lobelia, 8; *dortmanna* L., 89, 152, 157; *erinus* L., 89; *urens* L., 79, 89, 95
Lolium temulentum L., 191f
London, 14, 15, 196; Clay, 48; rocket, see *Sisymbrium irio*
lords-and-ladies, see *Arum maculatum*
lotus, 29
Lotus corniculatus L., 89, 99, **100**f; *tenuis* Waldst. & Kit., **100**f; *uliginosus* Schkuhr, **100**f
Lousley, J. E., 4, 96, 97, 198, 201, 210
lovage, see *Ligusticum scoticum*
Lower Greensand, 67
Lupinus nootkatensis Donn, 151, 163
Luronium natans (L.) Rafin., 162
Lusitanian distribution: a distribution along the extreme western seaboard of Europe, usually including Spain, Portugal and Western France, and frequently Ireland and the south-west of the British Isles, 87, 136f
Luzula arcuata (Wahl.) Wahl., 115, **123**f
Lychnis alba see *Silene alba*; *dioica* see *S. dioica*; *githago* see *Agrostemma githago*
Lycopodium alpinion L., 94; *annotinum* L., 94; *clavatum* L., 94; *inundatum* L., 94; *selago* L., 94
lyme grass, see *Elymus arenarius*
Lysaght, A., 4, 23
Lythrum salicaria L., 58, **76**f, 148
McLINTOCK, D., 209
McLean, R. C., 99, 210
Magnesian limestone, 99

Magnolia, 48
maiden pink, see *Dianthus deltoides*
Makins, F., 209
Malham, 97, 106, 108, 152
mallow, see *Malva*
Malva sylvestris L., 177
man, effect on vegetation, **54**, 57, 81, 97, 106, **110**f, 113, 142, 147, 151, 189
marjoram, see *Origanum*
marram grass, see *Ammophila arenaria*
Marsden-Jones, E. M., 26, 65, 180
marsh, 132, **146**f; bird's foot trefoil, see *Lotus uliginosus*; fern, see *Thelypteris palustris*; hawk's-beard, see *Crepis paludosa*; horsetail, see *Equisetum palustre*; mallow, see *Althaea officinalis*; orchid, see *Orchis latifolia*; pea, see *Lathyrus palustris*; sow-thistle, see *Sonchus palustris*; thistle, see *Cirsium palustre*; violet, see *Viola palustris*
Martyn, J., 15ff
Mat-grass, see *Nardus stricta*
Matricaria maritima L., subsp. *inodora* (L.) Clapham, *182*, 190, subsp. *maritima*, 190
Matthews, J. R., 55, 209
meadow saffron, see *Colchicum autumnale*
meadowsweet, see *Filipendula ulmaria*
Meconopsis cambrica (L.) Vig., *194*, 198
Medicago falcata L., 61
Meikle, R. D., 24
meiosis, 43
Melampyrum cristatum L., 13
melancholy thistle, see *Cirsium heterophyllum*
Melandrium album, see *Silene alba*; *dioicum* see *S. dioica*; *rubrum*, see *S. dioica*
melic grass, see *Melica*
Melica uniflora L., 71
Mendel, G., 26
Mentha, 79
Menyanthes trifoliata L., *135*, 149
Mercurialis perennis, L., 7, *35*, 59, 69, 72
Merrett, C., 11, 14
Mertensia maritima (L.) S. F. Gray, 172

Metcalfe, G., 87
mezereon, see *Daphne mexereon*
micro-climate, 37, 58, 77
microspecies, 122
military orchid, see *Orchis militaris*
milk parsley, see *Peucedanum palustre*
milk-thistle, see *Sonchus*
milk-vetch, see *Astragalus glycyphyllos*
Miller, P., 15
mimicry, 159
Mimulus, 163f; *guttatus* DC., 164; *luteus* L., 164
Minuartia stricta (Sw.) Hiern, 127
mint, see *Mentha*
mistletoe, see *Viscum album*
Molinia coerulea (L.) Moench, 83, 85f, 138
Moneses uniflora (L.) A. Gray, 74
monkey-flower, see *Mimulus*
Monocotyledons, 158, 159, 160, 161, 201
Montia, 149
Monotropa hypopitys L., agg., 68f
moonwort, see *Botrychium*
Morison, R., 14
moschatel, see *Adoxa*
moss (= bog), 133; campion, see *Silene acaulis*
Moss, C. E., 82, 209
mosses of snow-patches, 117f; of walls, 197
mossy saxifrage, see *Saxifraga hypnoides*
mountain avens, see *Dryas octopetala*; cranesbill, see *Geranium sylvaticum*; everlasting, see *Antennaria dioica*; limestone, see Carboniferous limestone; pansy, see *Viola lutea*; plants, adaptations of, **114**f, 118f, annual, 115; history of, 128f; on cliffs, 116, 124f; vegetative spread of, 115; with maritime distribution, 126f
sorrel, see *Oxyria digyna*
mouse-ear chickweed, see *Cerastium*; hawkweed, see *Hieracium pilosella*
mud crowfoot, see *Ranunculus lutarius*
mud-wort, see *Limosella*
Mulgedium alpinum, see *Cicerbita alpina*
mullein, see *Verbascum*

musk, see *Mimulus*
mustard, see *Sinapis alba*
mutation, 43
Mycelis muralis (L.) Rchb., 71, 219, 223
mycorrhiza: an assocation between a root and a fungus, 69, 74
Myosotis arvensis (L.) Hill, 65; var. *umbrosa* Bab., 65; *palustris* L., 148; *sylvatica* Ehrh., 65
Myrrhis odorata (L.) Scop., 213, 215
NAKED LADIES, see *Colchicum*
names of plants, after persons, 75; English, 73, 75, 86
Narcissus obvallaris Salisb., 185; *pseudonarcissus* L., *178*, 184f
Nardus stricta L., 83, 89, 124
Narthecium ossifragum L., 38, 140
Nasturtium sylvestre see *Rorippa sylvestris*
native plants, **110**f, 184, 190, 197f
Natural History Societies, local, 3f, 21, **23**f, 153, 161
Nature conservancy, 5, 77, 203
nature conservation, 5, 85, 147, 203; reserves, 136, 141, 143, 147
navelwort, see *Umbilicus*
nectary: an organ which secretes nectar, a sweet liquid exuded from various parts of many plants
Nelumbo nucifera Gaertn., 29
Neottia nidus-avis L., 68f
Nepta cataria L., 80; *glechoma* see *Glechoma hederacea*
nettle, see *Urtica*
nettle-leaved bellflower, see *Campanula trachelium*
new species, discovery of, in Britain, 27, 99f, 124
Newton, I., 12
Nidd, J., 12f
nipplewort, see *Lapsana*
Nuphar lutea (L.) Sm., 148, 159
Nymphaea alba L., *147*, 148, 149, 159
Nymphoides peltatum (Gmel.) O. Kuntze, 8, 13, 159
OAK, see *Quercus*
Obione portulacoides see *Halimione portulacoides*
ochrea: a sheathing stipule, typical of the family *Polygonaceae*, 168

Oenanthe aquatica (L.) Poir., 177; *crocata* L., 177; *fistulosa* L., 177; *fluviatilis* Coleman, 159, 177; *lachenalii* C. C. Gmel., 159, 177; *phellandrium* see *O. aquatica; pimpinelloides* L., 177; *silaifolia* Bieb., 177
Oenothera, 34
old man's beard, see *Clematis vitalba*
Oliver, F. W., 210
onion, see *Allium*
oolitic limestone, 64, 68, 70, 97, 101, 108, 187
open community (habitat): a community in which the plants do not grow close together and competition between them is light, 129, 189f
Ophioglossum vulgatum L., 8, 93f
Ophrys apifera Huds., 104
Opuntia, 191
orache, see *Atriplex*
Orchidaceae, 69, 104
Orchis incarnta see *O. strictifolia; latifolia* agg., 145; *mascula* L., *178; militaris* L., 4; *strictifolia* Opiz, 145, *146*
Origanum vulgare L., 70, 80
Orobanche minor Sm., *167*, 193
ovary, 34
ovate (leaf): egg-shaped, with the broader end at the base
ovule: a structure enclosing an ovum and developing after fertilization into a seed, 29, 34f
ovum: an egg, 29
Oxalis acetosella L., *66*, 68
ox-eye daisy, see *Chrysanthemum leucanthemum*
Oxford, 15f, 25, 62, 166, 182, 183, 197f; Botanic Garden, 198; Professors of Botany, 14, 16; ragwort, see *Senecio squalidus*
oxlip, see *Primula elatior;* false, 62
Oxycoccus macrocarpus (Ait.) Pers., 139
Oxyria digyna (L.) Hill, 125
oyster plant, see *Mertensia*
PALE BUTTERWORT, see *Pinguicula lusitanica*
pansy, wild, see *Viola arvensis* & *V. tricolor*
Papaver rhoeas L., *183*, 192

parasite (plant): a plant deriving its food from another living organism, to which it is attached, 193
Parey, A., 9
Parietaria diffusa Mert. & Koch, 197
Paris quadrifolia L., *35*, 64, 102
Parkinson, J., 10, 12
Parnassia palustris L., 146, *163*, 171
parsley piert, see *Aphanes*
pasque flower, see *Anemone pulsatilla*
Pastinaca sativa L., 104f, 177, 211, 215
Peace, T. R., 62
Pearsall, W. H., 113, 116, 210
peat, 49f, 84, 97, 129, **131**ff, 134, **140**f, 144, 146, 197
pedunculate oak, see *Quercus robur*
pellitory-of-the-wall, see *Parietaria diffusa*
Pennington, W., 56, 210
pennywort, see *Umbilicus*
pepper saxifrage, see *Silaum*
perennial plants, **35**, 167f, 171, 194, 196
perianth: the floral 'leaves' (calyx and corolla) considered as a unit
periglacial flora: the flora developed round the margin of an ice-sheet, 55
Perring, F. H., 4, 204, 208, 209
Petasites hybridus (L.) Gaertn., Mey. & Scherb., 161, 163
petiole: the stalk of a leaf
Petiver, J., 15
Petroselinum segetum (L.) Koch, 213, 215
Peucedanum palustre (L.) Moench, 143
photosynthesis: the formation in plant cells of food from water and carbon dioxide in the presence of chlorophyll and light
Phragmites communis Trin., 133, 141, 149, 153, 161
Picris echioides L., 219, 223; *hieracioides* L., 220, 223
pignut, see *Conopodium majus*
Pigott, C. D., 103
pill-wort, see *Pilularia*
Pilularia globulifera L., 155f

pimpernel, see *Anagallis*
Pimpinella magna see *P. major; major* (L.) Huds., 212, 216; *saxifraga* L., 105, 212, 216
pinewood, 54, **74**ff, 85, 114, 129
pin-eyed: a condition in primroses and other plants, where the stamens are shorter than the style, which alone can be seen in the throat of the flower, 76
Pinguicula alpina L., 136; *grandiflora* Lam., 136f; *lusitanica* L., *111*, 136; *vulgaris* L., 135f, 146
pinnate (leaf): composed of leaflets arranged in two rows along a stalk; in a 2-pinnate leaf the primary divisions are themselves pinnate; also 3-pinnate, etc.
Pinus sylvestris L., 50f, 52ff, **74**ff, 85, *98*, 114
plane, see *Platanus*
Plantago, 156; *maritima*, L., 26, 127, 176, 179
plantain, see *Plantago*
Platanus, 48
Plukenet, L., 15
Poa alpina L., 126; *annua* L., 45, 190
podsol; podsolised soil: a poor, acid soil with characteristic layering, due to soluble materials having been washed out of the surface-layers ('A' horizon) and deposited in the deeper layers ('B' horizon), 84, 116
poisonous plants, 70, **72**, 177, 185, 191f, 196
Polemonium coeruleum L., 108, 109
pollen analysis, **49**ff, 76, 84, 133; diagram, 49, *50*, 51; production, 49
pollination: the transfer of pollen from an anther to a stigma; cross, **33**, 34, 76f, 91, 126, 157; insect, **34**, 77, 88, 157; self, **33**, 34, 91, 135, 158; water, 174; water plants, 148, 174; without fertilization, 65
pollution of rivers, 163f
Polygonum hydropiper L., 152; *maritimum* L., 168, *raii* Bab., 168

polyploidy, **45**, 46, 101
Polytrichum norvegicum Hedw., 118
pond, 149, **151**ff, 156, 161, 164
pond-weed, see Potamogeton
poppy, see Papaver
Post Glacial period, 51f
Potamogeton, 35, 52, 159, 174
Potentilla fruticosa L., 127, 128; sterilis (L.) Garcke, 71
Powell, H. E., 8, 10
Praeger, R. L., 102
Price Evans, E., 210
prickly lettuce, see Lactuca serriola; pear, see Opuntia
Prime, C. T., 80, 209
primrose, see Primula vulgaris
Primula elatior (L.) Schreb., 61f; elatior x vulgaris, 61f; farinosa L., 146; veris L. 62, 178, 183; veris x vulgaris, 62; vulgaris Huds., 59, **60**f, 64, 76, 121
Prunella vulgaris L., 80
Psamma see Ammophila
Pteridium aquilinum (L.) Kuhn, 67, 74, **83**, 84, 93
Puccinellia maritima (Huds.) Parl., 175, 177
Pugsley, H. W., 18, 186, 201
Pulicaria dysenterica (L.) Bernh., 146
purple loosestrife, see Lythrum salicaria; milk-vetch, see Astragalus danicus; moor-grass, see Molinia coerulea
pyramidal orchid, see Anacamptis pyramidalis
Pyrola media Sw., 75; minor L., 75; rotundifolia L., 75; subsp. maritima (Kenyon) E. F. Warburg, 75, 171; secunda see Ramischia secunda; uniflora see Moneses
QUAKING GRASS, see Briza
Quaternary period, 48f
Quercus, 38, 40f, 53, 76, 129; petraea (Mattushka) Liebl., 58f, **66**f, 82, 113f; robur L., 38, **58**f, 66f; sessiliflora see Q petraea
quillwort, see Isoetes
Radiola linoides Roth, 90
ragwort, see Senecio jacobaea
Ramischia secunda (L.) Garcke, 75
ramsons, see Allium ursinum
Rand, I., 15, 16
Ranunculaceae, 63f, 65, 72

Ranunculus, 40f, **44**, 63f, 162, 183, 183; aquatilis L., 150; auricomus L., 65; circinatus Sibth., 150f; ficaria L., 51, 59, 60, 72; flammula L., 146; fluitans Lam., 150; glacialis L., 124; hederaceus L., 150f; lutarius (Rével) Bouvet, 150; parviflorus L., 30, 31; repens L., 146; subgenus Batrachium, 150, 150f
Raphanus maritimus Sm., 168; raphanistrum L., 168, 194
rare plants, Breckland, **84**f; conservation, 145, 203; fen, **143**ff; limestone cliffs, **108**f; Limonium, 180; quarries, 196; Wolffia, 155
Raunkiaer, C., 35
Raven, C. E., 1, 9, 11f, 98, 108, 210
Raven, J., 210
ray florets: the outer florets in the flower-head of a composite, usually markedly different from the central, disk florets
Ray, J., 1, 10ff, 16, 17, 19, 20, 84, 108, 168
receptacle: the modified upper part of a stem which bears the parts of a flower
red campion, see Silene dioica; clover, see Trifolium pratense; dead-nettle, see Lamium purpureum
reed-mace, see Typha
reed-swamp, **141**f, 144, 149, 152, 161
Reid, E. M., 48
Rhamnus cathartica L., 76; frangula see Frangula alnus
rhizome: an underground stem of rootlike appearance
Rhododendron, 87; ponticum L., 111
Richardson, R., 16
Riddelsdell, H. J., et al., 22
Ridley, N., 7
Rishbeth, J., 197
rock-rose, see Helianthemum
rootstock, see rhizome
Rorippa sylvestris (L.) Besser, 196
Rosa spinosissima L., 170
rose bay willow-herb, see Chamaenerion angustifolium
rose-root, see Sedum rosea
Ross-Craig, S., 209

Rothiemurchus Forest, 114
rough chervil, see Chaerophyllum temulum; hawkbit, see Leontodon hispidus; hawks-beard, see Crepis biennis
Rubus fruticosus agg., 67, 79, 122
Rudbeckia, 32
Rumex, 45, 201; crispus L., var. trigranulatus Syme, 172
rush, see Juncaceae and Juncus
Russell, B., 2
SAFFRON, see Crocus sativus
sage, see Salvia
Sagittaria sagittifolia L., 147, 162
St. John's-wort, see Hypericum
St. Vincent's Rocks, 8
Salicornia, 167, **174**f, 179
Salisbury, E. J., 31, 70f, 169 180, 190, 199, 210
Salisbury, R. A., 123
Salix, 29, 30; herbacea L., 52, 117f; lanata L., 126; repens L., 145, 171
Salsola kali L., 166f
salt, influence on vegetation, 166, 173, 178; marsh, 132, **173**ff
saltwort, see Salsola kali
Salvia cleistogama de Bary & Paul, 34; horminoides Pourr., 80; verbenaca see S. horminoides
Samolus valerandi L., 145
samphire, see Crithmum maritimum, Inula crithmoides, and Salicornia
sand sedge, see Carex arenaria; spurrey, see Spergularia
sand-dune, 74, 75, 107, 145, **166**ff, 190
sandwort, see Arenaria, Minuartia and Honkenya
Sanguisorba officinalis L., 146
Sanicula europaea L., 69
saprophyte: a plant deriving its food from dead organic matter, 68f, 74
Sarothamnus scoparius (L.) Wimmer, 82, 88
Saussurea alpina (L.) DC., 125
Savidge, J. P., 204
Saxifraga aizoides L., 121; cernua L., 124; hypnoides L., 125; oppositifolia L., 14, 116, 120f, 125; stellaris L., 116ff, 125, 131
Scandix pecten-veneris L., 212, 216

scarlet pimpernel, see *Anagallis arvensis*
Scheuchzeria palustris L., 140
Schoenoplectus lacustris (L.) Palla, 139, 150
Schotsman, H. D., 163
Scilla nonscripta see *Endymion nonscriptus*; *nutans* see *Endymion nonscriptus*; *autumnalis* L., 180; *verna* Huds., 180
Scirpus caespitosus see *Trichophorum caespitosum*; *lacustris* see *Schoenoplectus lacustris*
Scots pine, see *Pinus sylvestris*
sea beet, see *Beta maritima*; buckthorn, see *Hippophaë rhamnoides*; cabbage, see *Brassica oleracea*; campion, see *Silene maritima*; carrot, see *Daucus gingidium*; cliffs, 178ff, 197; couch grass, see *Agropyron junceiforme*; holly, see *Eryngium maritimum*; kale, see *Crambe maritima*; lavender, see *Limonium*; lungwort, see *Mertensia maritima*; pea, see *Lathyrus japonicus*; pink, see *Armeria maritima*; purslane, see *Halimione portulacoides*; radish, see *Raphanus maritimus*; rocket, see *Cakile maritima*; sandwort, see *Honkenya peploides*; spurge, see *Euphorbia paralias*; stork's-bill, see *Erodium maritimum*
sedge, see *Carex*; fenman's, see *Cladium*
Sedum, 197; 176; *acre* L., 176; *rhodiola* see *S. rosea*; *rosea* (L.) Scop., 103, 125
seed, dormancy, 29f, 192; germination, 29ff, 169, 180, 197, 203; output 29, 32, 180, 193; purity of agricultural, 191f, 193; weight, 70f
Selinum carvifolia L., 143
self-heal, see *Prunella vulgaris*
Sell, P. D., 4, 209
Senebiera coronopus see *Coronopus squamatus*; *didyma* see *Coronopus didymus*
Senecio jacobaea L., 172; *squalidus* L., 194, 195, 198; *vulgaris* L., 190, 191, 194
sens. lat., see agg.
sens. strict (= sensu stricto), after a name means that it is used in a narrow sense (cf. agg.)

sere, 39
serendipity nicely balanced by hope, 4
setaceous: bristle-like
Seseli libanotis (L.) Koch, 105
Sesleria coerulea (L.) Ard., 106f
sessile oak, see *Quercus petraea*
sheep's fescue, see *Festuca ovina*
shepherd's needle, see *Scandix pecten-veneris*; purse, see *Capsella bursa-pastoris*
Sherard, W., 16
shrubby cinquefoil, see *Potentilla fruticosa*; sea-blite, see *Suaeda fruticosa*
Silaum silaus (L.) Schinz & Thell., 143, 211, 216
Silene acaulis L., 114f, 125, 131; *alba* (Mill.) Krause, 45, 190; *conica* L., 31f; *cucubalus*, see *S. vulgaris*; *dioica* (L.) Clairv., 45, 61, 73, 124, 190; *maritima* With., 127, 129, 172, 179; *vulgaris* (Moench) Garcke 179f, 204
silver fir, see *Abies alba*
Simpson, N. D., 18, 208
Sinapis alba L., 158, 194; *arvensis* L., 192, 194
Sison amomum L., 213, 216
Sisymbrium irio L., 196; *officinale* (L.) Scop., 195; *orientale* L., 195f; *thalianum* see *Aralidopsis thaliana*
slender bird's foot trefoil, see *Lotus tenuis*
Sloane, H., 15
small flote-grass, see *Glyceria declinata*; hare's ear, see *Bupleurum opacum*
Smith, J. E., 19f
Smith W. G., 209
smooth cat's ear, see *Hypochaeris glabra*; hawk's-beard see *Crepis capillaris*
Smyrnium olusatrum, L., 211, 216
snake's head, see *Fritillaria*
snow buttercup, see *Ranunculus glacialis*; gentian, see *Gentiana nivalis*
Society for the Promotion of Nature Reserves, 5
soft hawk's-beard, see *Crepis mollis*
Solanum dulcamara L., 76, 173, 196; var. *marinum* Bab., 173
Soldanella, 119

Solidago cambrica Huds., 68; *virgaurea* L., 67
Sonchus arvensis L., 172, 220, 223; *asper* (L.) Hill, 220, 223; *oleraceus* L., 172, 220, 223; *palustris* L., 220, 223
Sorbus, 56; *aria* L., 109; *minima* (Ley) Hedl., 110; *rupicola* (Syme) Hedl., 110
sow-thistle, see *Sonchus*
Sowerby, James, 19
sp. = species (singular)
Sparganium ramosum Huds., 150
Spartina alterniflora Lois., 174; *maritima* (Curt.) Fernald, 174f; *stricta* see *S. maritima*; *townsendii* H. & J. Groves, 45, 174f
speciation: the formation of species during evolution, 44f
species, 40ff *et passim*
speedwell, see *Veronica*
Spergula arvensis L., 189
Spergularia marina (L.) Griseb., 175f; *media* (L.) C. Presl., 175f, 178
Sphagnum, 49, 93, 131, 132f, 135, 136, 139, 146
spiny milk-thistle, see *Sonchus asper*; sow-thistle, see *Sonchus asper*
Spiraea ulmaria see *Filipendula ulmaria*
Spiranthes spiralis (L.) Chevall., 181
spleenwort, see *Asplenium*
spore: a small, simple, asexual reproductive body, 156, 157
spotted cat's-ear, see *Hypochaeris maculata*
spp. = species (plural)
spreading hedge parsley, see *Torilis arvensis*
spurge laurel, see *Daphne laureola*
spurrey, see *Spergula arvensis*
squill, see *Scilla*
squinancy-wort, see *Asperula cynanchica*
Stachys officinalis (L.) Trev., 39, 80; *sylvatica* L., 58, 80
stag's horn club-moss, see *Lycopodium clavatum*
starry saxifrage, see *Saxifraga stellaris*
Statice, see *Limonium*
Stellaria, 75; *media* L., 190, 191
Step, E., 209

240INDEX

stinking hawk's-beard, see *Crepis foetida*; hellebore, see *Helleborus foetidus*
stipule: a leaf or scale-like organ at the base of a petiole
stone parsley, see *Sison amomum* and *Seseli libanotis*
stonecrop, see *Sedum*
Stonehouse, W., 10, 14
stork's-bill, see *Erodium*
Stratiotes aloides L., 160, 162
strawberry, see *Fragaria*; clover, see *Trifolium fragiferum*; origin of garden, 71; tree, see *Arbutus*
Strudwick, F. E., 209
Suaeda fruticosa Forsk., 172, 175; *maritima* (L.) Dum., 175
Sub-Atlantic period, **54**, 56 129, 134
Sub-Boreal period, **54**, 134
sub-fossil plant remains, **49**f, 190
subsp. = subspecies
Subularia aquatica L., 158
subulate: awl-shaped, i.e., narrow and pointed
Succisa pratensis Moench, 66, 125
Summerhayes, V. S., 181, 210
sundew, see *Drosera*
sweet cicely, see *Myrrhis odorata*; flag, see *Acorus calamus*; peas, 30; violet, see *Viola odorata*
swine's cress, see *Coronopus*; succory, see *Arnoseris minima*
Syme, J. T. J. B., 22, 209
Symphytum, 204
TANSLEY, A. G., 26, 38, 39, 40, 76, 173, 208, 210
Taraxacum, 122, 191, 218, 224
Taxon (pl. taxa): any taxonomic category, e.g., a variety, species, genus, family, etc.
Taxonomy: the science of classification; (in biology) the classification of plants and animals into taxa, i.e., the categories embodied in the International Codes of Botanical and Zoological Nomenclature, 17, **26**f, 204
Taxonomy, experimental, **25**ff, 203f
Taxus baccata, L., 109

teasel, see *Dipsacus*
Teesdale, 21, 93, 99, 102, 107, **127**, 129, 130
Teesdalia nudicaulis (L.) R.Br., 85
Tenby daffodil, see *Narcissus obvallaris*
ternate (leaf): divided into three approx. equal parts, each part being a single leaflet (1-ternate), or itself ternate (2-ternate); similarly 3-ternate
Tertiary period, 48
tetraploid, **45**, 101, 163, 184
Teucrium botrys L., 67; *chamaedrys* L., 67; *scordium* L., 67, *134*; *scorodonia* L., 67
thale cress, see *Arabidopsis*
Thelypteris palustris Schott, 145
Thesium humifusum DC., 13
thorow-wax, see *Bupleurum rotundifolium*
Thrincia hirta see *Leontodon leysseri*
thrum-eyed: a condition, in primroses and other plants, where the style is shorter than the stamens which alone can be seen in the throat of the flower, 76
thyme, see *Thymus*
thyme-leaved speedwell, see *Veronica serpyllifolia*
Thymus, 120; *drucei* Ronn., *99*, 103, 106, 110; *pulegioides* L., 103; *serpyllum* L., 103
Tilia, 53, 129
Tillaea muscosa L., 90
Torilis anthriscus see *T. japonica*; *arvensis* (Huds.) Link, 212, 217; *infesta* see *T. arvensis*; *japonica* (Houtt.) DC., 79, 213, 217; *nodosa* (L.) Gaertn., 212, 217
Tournefort, J. P., 16, 179
tower-cress, see *Arabis turrita*
Tragopogon pratensis L., *38*, 219, 224
transplant experiments, 26
Trapa natans L., 48
traveller's joy, see *Clematis*
tree mallow, see *Lavatera*
trefoil, see *Trifolium*
Trichophorum caespitosum (L.) Hartm., 82, 139
Trientalis europaea L., 75
Trifolium fragiferum L., *166*, 177; *maritimum* see *T.*

squamosum; pratense L., 193; *squamosum* L., 177
Triglochin, 140
Trimen, H., 15, 22
trimorphous: having three forms
Trinia glauca (L.) Dum., 105
triploid, 162
Trollius europaeus L., 125, 126, 146
tuber: a swollen portion of a root or stem, usually underground
tufted hair-grass, see *Deschampsia caespitosa*; vetch, see *Vicia cracca*
Tulipa australis Link, 184; *sylvestris* L., 183f
tundra, 52, 53, 114, 128
Turesson, G., 26, 67
Turner, W., 7, 14, 27, 72
Turrill, W. B., 26, 28, 180, 204, 208, 210
Tussilago farfara L., *191*, 194
Tutin, T. G., xiii, xiv, 201, 207-9
Typha, 161
twayblade, see *Listera ovata*
Ulex, 86, 88, 95, 193; *europaeus* L., *83*, 88; *gallii* Planch., 88; *minor* Roth, 88
Ulmus, 53, 129
umbel: an inflorescence in which the branches all arise from approximately one point
Umbelliferae, 79, 104f, 143, 159, 170, 177, 179, 186, 211ff
Umbilicus rupestris (Salisb.) Dandy, 197
upright brome, see *Zerna erecta*; hedge parsley, see *Torilis japonica*
Urtica, 193
Utricularia intermedia Hayne, 137; *vulgaris* L., 135, 137, 146
Vaccinium x *intermedium* Ruthe, 88; *myrtillus* L., 67, 74, **87**f, 95, 124, 134; *uliginosum* L., 88; *vitis-idaea* L., 87
Valentine, D. H., 61, 161
vegetative reproduction, *Anemone nemorosa*, 60; *Chamaenerion*, 194; *Circaea*, 66; dune plants, 168f; *Lotus corniculatus*, 100; *Lycopodium selago*, 94; mountain plants, 115; *Paris*, 64; *Pterdium*, 93; *Tulipa*, 184; waterplants, 149, 153f, 160; weeds, 191

Verbascum nigrum L., 187; *thapsus* L., 30, 32, *179*, 187
vernal squill, see *Scilla verna*
Veronica, 84; *agrestis* L., 192; *buxbaumii* see *V. persica*; *hederifolia* L., 30, 31; *humifusa* Dicks., 126; *persica* Poir., 190, 192f; *polita* Fries, 192f; *praecox* All., 85, *174*; *serpyllifolia* L., 126; *spicata* L., 85, *102*; *triphyllos* L., 85
Viburnum opulus L., 76
Vicia cracca L., 78; *sepium* L., 78
Villarsia nymphaeoides see *Nymphoides peltatum*
Vinca major L., 17; *minor* L., 17
Viola, 43, 90ff; *arvensis* Murr., 107; *canina* L., 91ff; subsp. *montana* (L.) Fries, 93; var. *ericetorum* Rchb., 92; x *V. riviniana*, 91f; *curtisii* (Forst.) Syme, 107; *hirta* L., 92; *lactea* Sm., 91f; *laceta* x *V. riviniana*, 91f; *lutea* Huds., 107; *odorata* L., 34, 90, 92; *palustris* L., 93, 146; *reichenbachiana* Jord., 63, *92*; *riviniana* Rchb., 63, *67*, 90ff, 124; subsp. *minor* (Murb.) Valentine, 90; subsp. *nemorosa* Neum., 90 *rupestris* Schmidt, 93, 99f, 127, 128; *stagnina* Kit., 93, 144; *sylvestris* see *V. reichenbachiana*, *tricolor* L., 107, *114*
violet, see *Viola*
Viscum album L., 70
von Post, L., 51
Wahlenbergia hederacea (L.) Rchb., 89
wall lettuce, see *Mycelis muralis*; rocket, see *Diplotaxis muralis*
Wallace, E. C., 99
wallflower, see *Cheiranthus cheiri*
Walters, S. M., 128, 151, 209, 210
Warburg, E. F., xiii, xiv, 207-9
Warming, J. E. B., 26
Warner, R., 22
water avens, see *Geum rivale*; crowfoot, see *Ranunculus* subgenus *Batrachium*; dispersal, 152f; ferns, 155f;

forget-me-not, see *Myosotis palustris*; germander, see *Teucrium scordium*; impermanence of habitats, 151f; lobelia, see *Lobelia dortmanna*; pepper, see *Polygonum hydropiper*; soldier, see *Stratiotes aloides*; uniformity of habitats, 154; violet, see *Hottonia palustris*
water plants, distribution, 153; in typically 'land' families, 158f; irregular flowering and fruiting 149; over-wintering, 155, 160; plasticity, 150f, 162; pollination, 148, 157f; structural features, 149, 156
water-blinks, see *Montia*
water-chestnut, see *Trapa*
water-lily, see *Nymphaea* and *Nuphar*
Watson, H. C., 4, **24**, 209
Watt, A. S., 84, 94
wavy hair grass, see *Deschampsia flexuosa*
weeds, 53, 84, 111, 186, **189**ff
Welsh poppy, see *Meconopsis cambrica*
Went, F. W., 33
white campion, see *Silene alba*; helleborine, see *Cephalanthera damasonium*; water-lily, see *Nymphaea alba*
White, J. W., 22, 73, 179
whitebeam, see *Sorbus*
whitlow-grass, see *Erophila verna*
whortleberry, see *Vaccinium myrtillus*
Wicken Fen, 137f, **141**ff, 158
wild angelica, see *Angelica sylvestris*; basil, see *Clinopodium vulgare*; carrot, see *Daucus carota*; clary, see *Salvia horminoides*; daffodil, see *Narcissus pseudonarcissus*; garlic, see *Allium ursinum*; hyacinth, see *Endymion nonscriptus*; iris, see *Iris pseudacorus*; lettuce, see *Lactuca virosa*; pansy, see *Viola arvensis* and *V. tricolor*; parsnip, see *Pastinaca sativa*; strawberry, see *Fragaria vesca*; tulip, see *Tulipa sylvestris*
Wild Flower Society, **4**

Willis, J. C., 210
Willisel, T., 11
willow, see *Salix*
willow-herb, see *Epilobium* and *Chamaenerion*
Willughby, F., 13
Wilmott, A. J., *18*
Wilson, J., 17
windflower, see *Anemone nemorosa*
winter aconite, see *Eranthis hiemalis*; annual, 30, 85; buds, 155, 160
winter-cress, see *Barbarea vulgaris*
wintergreen, 74f
Wolffia arrhiza (L.) Wimm., 155
wood anemone, see *Anemone nemorosa*; avens, see *Geum urbanum*; sanicle, see *Sanicula europaea*; sorrel, see *Oxalis acetosella*
Woodell, S. R. J., 61
Woodhead, N., 123, 157
woodruff, see *Asperula odorata*
wood-sage, see *Teucrium scorodonia*
woodspurge, see *Euphorbia amygdaloides*
woody nightshade, see *Solanum dulcamara*
woolly-headed thistle, see *Cirsium eriophorum*
work, suggestions for further, 2f, 200ff; Batrachian *Ranunculi*, 150; colonisation of ponds, 153; ecology of rare plants, 109; *Gagea*, 63; *Impatiens glandulifera*, 164; *Lobelia dortmanna*, 157; *Primula*, 61; sea-cliff vegetation, 178; snow-patch vegetation, 119; *Viola*, 91f; water-plants, 164f
YELLOW ARCHANGEL, see *Galeobdolon luteum*; bird's nest, see *Monotropa*; deadnettle, see *Galeobdolon luteum*; iris, see *Iris pseudacorus*; rocket, see *Barbarea vulgaris*; saxifrage, see *Saxifraga aizoides*; toadflax, see *Linaria vulgaris*; waterlily, see *Nuphar lutea*
yew, see *Taxus baccata*
Young, D. P., 69, 201
Youngman, B. J., 29
Zerna erecta (Huds.) Panz., 98, 99
Zostera, 174